D0793968

SAP Excellence

Series Editors:
Professor Dr. Dr. h.c. mult. Peter Mertens
Universität Erlangen-Nürnberg

Dr. Peter Zencke
SAP AG, Walldorf

Gerhard F. Knolmayer, Peter Mertens
Alexander Zeier and Jörg Thomas
Dickersbach

Supply Chain Management Based on SAP Systems

Architecture and Planning Processes

With 77 Figures
and 11 Tables

 Springer

Prof. Dr. Gerhard F. Knolmayer
University of Bern
Institute of Information Systems
Engehaldenstrasse 8
CH-3012 Bern
Switzerland

Prof. Dr. Dr. h.c. mult. Peter Mertens
University of Erlangen-Nürnberg
Department of Information Systems I
Lange Gasse 20
D-90403 Nürnberg
Germany

Dr. Alexander Zeier
Deputy Professor of
Prof. Hasso Plattner
Hasso Plattner Institute
University of Potsdam
August-Bebel-Str. 88
D-14482 Potsdam
Germany

Dr. Jörg Thomas Dickersbach
Von-der-Pfordten-Str. 18
D-80687 Munich
Germany

"SAP", "SAP AII" "SAP APO", "SAP BI", "SAP CRM", "SAP ECC" "SAP EM", "SAP ERM", "SAP ERP", "SAP EWM", "SAP ICH", "SAP EP", "SAP F&R", "SAP NetWeaver", "SAP OER", "SAP R/2", "SAP R/3", "SAP SCM", "SAP SEM", "SAP SNC", "SAP SRM", "SAP Inventory Collaboration Hub", "SAP Business ByDesign", and "SAP XI" are trademarks of SAP Aktiengesellschaft, Dietmar-Hopp-Allee 16, D-69190 Walldorf, Germany. The publisher gratefully acknowledges SAP's kind permission to use its trademark in the publication. SAP AG is not the publisher of this book and is not responsible for it under any aspect of press law.

ISBN 978-3-540-68737-5   e-ISBN 978-3-540-68739-9

DOI 10.1007/978-3-540-68739-9

Library of Congress Control Number: 2008936141

© 2009 Springer-Verlag Berlin Heidelberg

Cover design: WMX Design GmbH, Heidelberg

Printed on acid-free paper

9 8 7 6 5 4 3 2 1

Springer.com

Foreword by Prof. Hasso Plattner

The accelerating pace of globalization today poses increased challenges for Supply Chain Management (SCM). To offer just one product to the customer, sometimes hundreds of companies must collaborate along the value chain. Understanding and optimizing this logistical complexity can be a competitive advantage if it is handled effectively with state-of-the-art IT systems. This is why SAP has been intensively engaged in developing powerful SCM applications for about a decade.

Since SAP SCM is one of the most challenging systems in our portfolio, we have turned to the best experts we know on the subject matter for insight – our customers. They articulated industry-specific needs and challenged our development teams with their visions of real-time enterprises and market-driven supply chains. Many powerful features have been developed in close cooperation with these customers.

SAP has responded by assigning some of its most talented system architects and programmers to SCM development. To ensure that these solutions work in the real world, SAP has devoted significant resources towards building components and integrating them into large, realistic system landscapes. For instance, we have invested a great deal of effort in achieving short response times, which are crucial for customers' acceptance of IT systems, even in the case of very large data volumes.

The authors of this book combine the latest research findings and practical experience with SCM systems striking a balance between managerial overview and practical detail. I especially appreciate the perspective on SAP's most recent development strategy, which enables Small and Medium Enterprises (SMEs) to service-enable their business processes, leveraging the ecosystem provided through SAP. The new component-based application platform together with the principles of SAP's Enterprise Service-Oriented Architecture provides the necessary foundation for the next generation of products. Our newest product, SAP Business ByDesign, underlines the paradigm shift in enterprise software towards Software-as-a-Service (SaaS), which will increase the competitiveness of SMEs by allowing smaller and smaller companies to participate in global value chains.

Prof. Dr. h.c. mult. Hasso Plattner
Chairman of the Supervisory Board and Co-Founder, SAP AG

Preface

Since the publication of the first version of this highly successful book in 2000 it has become obvious that Supply Chain Management (SCM) is not just a buzzword or a fashion, as are so many other subjects in management and IT. Some "mega-trends" in the world of business and economy, such as globalization and growing technical specialization of product and process development, result in a complex division of labor at national and international levels and in reduced net value added within single firms.

To coordinate all these elements of development, procurement, production, sales, distribution, and recycling functions, SCM has become a mandatory approach. Successful SCM must be based on powerful IT systems that combine traditional ERP systems with specific SCM functionality, including sophisticated Operations Research heuristics and algorithms.

With respect to their leading position in the market, SAP systems are highly relevant in the management of supply chains. Because of the extensive progress of SAP SCM™ systems in recent years, the English book published in 2002 has had to be totally rewritten. The text was finished in spring 2008 and refers to SAP SCM™ 5.0.

Gerhard Knolmayer, Peter Mertens, and Alexander Zeier, who authored earlier versions of the book, are deeply grateful to Jörg Thomas Dickersbach, who has joined the authoring team and brought with him his profound knowledge of and detailed experience with SAP SCM™ systems.

The book has also gained from our cooperation with several top SCM specialists working in globally operating companies such as Colgate-Palmolive, Danfoss, Henkel, Hilti, and Nestlé, who describe the implementation, usage, and experiences with SAP SCM™ systems and share their views on the future of SCM in Chapter 5.

This book is also among the first to discuss the impact of SAP's new Business ByDesign System on SCM™. We are much indebted to Wilhelm Zwerger and Dr. Bülent Akinto of SAP for contributing this information in Chapter 6 of the book. We also appreciate different types of support from members of our teams, in particular Dr. Dina Barbian, Lukas Helfenstein, Gabriela Loosli, Jürgen Müller, Manuela Stolz, Daniel Stucki, and Thomas Wermelinger.

Gerhard F. Knolmayer, Bern
Peter Mertens, Erlangen-Nürnberg
Alexander Zeier, Potsdam
Jörg Thomas Dickersbach, Munich

Table of Contents

Chapter 1

Introduction

In the last decade, in conjunction with extensive reorganizations of business structures and processes, the concept of "Supply Chain Management" (SCM) has
gained rapidly importance. Many companies have realized reengineering projects;
they often reduced the degree of vertical integration, thus obtaining ever more
products and services from external suppliers. With concepts such as virtual companies, extended enterprises, strategic alliances, and company networks, the legal
and business limits of companies are becoming blurred. Consequently, the coordination of business processes beyond the elementary organization units gains
particularly in importance. Whereas "lean management" tries to counter various
forms of waste within a company, SCM aims at avoiding waste all along the value
chain (cf. Plenert 2007). And the turbulent changes in economic environments ask
for agile supply chains and real-time transfer at least of selected data.

The more companies are involved in producing services and products, the better
they can concentrate on their core competencies. This, however, also increases a
number of interfaces between them. Consequently, overall planning, scheduling,
monitoring, and controlling activities of inter-company processes gain in importance.

Problems resulting from poor SCM, such as production or shipment delays,
may have a severe impact on the market value of a company: A large event study
has shown that SCM-related glitch announcements result in an abnormal decrease
in shareholder value by more than 10%. An example of this effect is shown in
Fig. 1.1 (Hendricks and Singhal 2003; Singhal 2003).

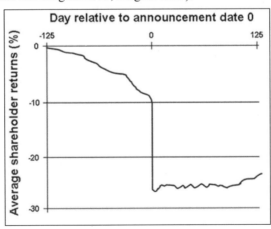

Fig. 1.1 Stock quotation before and after announcement of an SCM glitch

An impression of the importance of SCM in our global economy can be gained from the two following examples from different industries:

High-technology parts for power generation gas turbines are produced along a 5,000-mile-long supply chain. A blade for a gas turbine may be cast in the United States, transported from the east coast to England for machining, to Italy for coating, for laser hole drilling back to England, and for rotor assembly to Switzerland. The rotor may be transported to Germany for assembly with stationary parts of the turbine. The assembled product, weighting about 150 tons or 300,000 lb, may be transported by ship down the Rhine and reloaded onto a seagoing vessel for transport to the customer.

The development and production of a personal computer hard disk typically also involves several continents (de Souza and Khong 1999). A Seagate disk may be developed in the United States, and the wafer for the read/write heads in Northern Ireland. The heads are cut out of the wafers in Malaysia or Thailand and mounted on metal-arms in Thailand or China. A Japanese supplier produces aluminum disks, which are prepared for coating in Northern Ireland. Coating is done in California or Singapore. The final assembly takes place in China, Thailand, or Singapore. And the supply chain may change if new products or production processes are introduced.

In the last decade, the theory of SCM has advanced in many directions. Logistical issues have been reformulated under an SCM perspective; the need to consider environmental aspects has become obvious; risk management in supply chains has gained much interest; contracts have been analyzed as a means of risk sharing; and the effects of such major trends as globalization and outsourcing have been studied from viewpoints won through game theory, with the role of information technology (IT) in SCM becoming accentuated even more than before.

SCM would not be possible without the advances in IT and information systems (IS). Unfortunately, the two strands of conceptual research and research on the present state of IT systems for SCM remain largely unconnected. For instance, the six perspectives on SCM formulated by Otto and Kotzab (2003) do not even mention IT systems. In this book we try to close the gap that exists between the issues regarded as relevant for successful SCM and the support IT systems are offering for SCM today.

Just 10 years ago, specialized vendors of SCM systems were at the forefront of SCM innovation. Today, most companies running SCM systems rely on the vendor of an ERP system that is already installed (cf. the Hilti and Nestlé case studies in Chapter 5). Given the market shares, it makes sense to focus on the SCM offerings of SAP AG and compare the state of its SAP SCM™ system (as of Release 5.0) with requirements for a good IT support of SCM.

The remainder of this book is organized as follows: Chapter 2 discusses the scope of SCM, focusing on collaboration issues, and defines requirements and wishes that we derived from our study of the research literature on SCM and from our discussions with SCM managers. We propose an SCM pyramid to structure the tasks related to different levels of SCM and discuss the impact of industry-specific issues on IT support for SCM. In Chapter 3 we give a short overview of the process landscape as seen by SAP. This process map is further detailed in Chapter 4, which provides a description of selected functionalities available in SAP's SCM applications:

- SAP Advanced Planning and Optimization (SAP APO™),
- SAP Forecasting and Replenishment (SAP F&R™)
- SAP Inventory Collaboration Hub (SAP ICH™), recently renamed as SAP Supply Network Collaboration (SAP SNC™)
- SAP Event Management (SAP EM™).

In this book we will not deal with SAP's fifth SCM application, the SAP Extended Warehouse Management (SAP EWM™).

A comparison of SCM functionalities from different viewpoints shows that a broad spectrum of business requirements is already covered by SAP's SCM offerings, but also that some areas exist that are still uncovered and in which further improvements may be worthwhile. In Chapter 5 we describe experiences of several firms during their implementation and using of SAP SCM™ systems; we are highly appreciative that the following companies contributed information for this section:

- Colgate-Palmolive
- Danfoss
- Henkel
- Hilti
- Nestlé.

In Chapter 6 we pick up a very recent development in the SCM market and provide first-hand information on SAP's new Business ByDesign™ product and its perspectives for supporting SCM in medium-sized companies.

1.1   Definitions and Terminology

Supply Chain Management tries to improve the flow of

- materials,
- information, and
- financial resources

within the company and among companies collaborating under long- or medium-term agreements by

- sharing information,
- concerted planning and scheduling,
- coordinated execution, and
- collaborative monitoring and controlling

to improve the competitiveness of the entire supply chain.

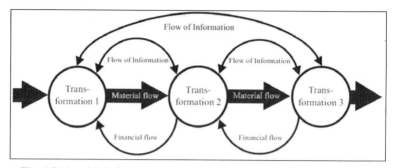

Fig. 1.2 Material, information, and financial flows as core elements of SCM

Fig. 1.2 shows the typical directions of flows; however, in the case of return shipments or refunds different flow directions result.

Supply chains may consist of independent companies, but can also be made up of organizational entities that legally belong to one group. Whereas the literature concentrates on inter-company collaboration, most SCM projects in practice concern intra-group systems and collaboration within a group; the case studies in Chapter 5 confirm this.

The constitutive term "chain," on which SCM is based, provides an incorrect view of business realities. Typically, many business relationships are relevant in producing a certain product or service. The idea of a "Supply Network," a "Supply Web," a "Value Net," or a logistics network would be more appropriate, because a company typically belongs to several supply chains (cf. Fig. 1.3). However, only the main business relationships demand close collaboration, and the partnering companies should be selected with much care. For less important supplies, short-term, unstable procurement decisions that result from auctions may be more adequate. In this book we do not refer to such relationships as "Supply Chains," and we refer to SCM also in network systems.

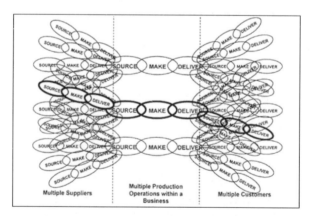

Fig. 1.3 Supply chain as part of a supply network
(based on earlier material from the Supply-Chain Council)

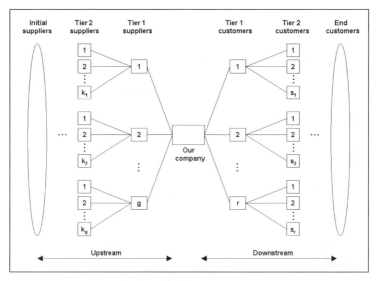

Fig. 1.4 Visualization of different tiers in the supply chain

From a certain position in the supply chain it is possible to look "upstream" (toward suppliers) or "downstream" (toward customers). Thus, SCM encompasses not only the supply side, but also the demand side is subject of SCM. Depending on the "distance" from the company considered, tier 1, tier 2, up to "tier n" suppliers and customers are distinguished (Fig. 1.4).

1.2   Benefits of Supply Chain Management

In the definition of SCM, we emphasized that it should make the supply chain more competitive. SCM may be viewed as an approach to improving the long-term competitive position and short-term profitability of the collaborating companies. To improve the competitive position, the entities of the supply chain could, for example,

- coordinate investments,
- share the use of resources,
- avoid redundancies, e.g., in quality control,
- collaborate in R&D and forecasting,
- incorporate customers and suppliers in design processes and value analyses,
- redesign structures and processes, and
- reposition functionalities (e.g., managing inventories) between the partners.

Profitability may be raised by higher revenues and/or lower costs. The widely recognized DuPont scheme for explaining the Return on Investment (ROI = profit/ capital) shows that SCM has several points to contribute to a higher ROI (cf. Schnetzler and Schönsleben 2007):

- Higher profit (i.e., increasing the numerator of the ROI formula)
  - by realizing higher revenues, e.g.,
    - by reducing time-to-market by better coordinated design processes, or
    - by better service levels resulting from repositioning inventory;
  - by selling products with bigger margins, made possible (e.g.) by more cost-efficient procurement, manufacturing, and logistics processes.
- Reducing locked assets and capital (i.e., decreasing the denominator of the ROI formula), e.g.,
  - by reducing inventory or fixed assets by more efficient procurement, production, and distribution planning or
  - by shortening the cash-to-cash cycle time.

From another perspective, the benefits of SCM systems can be categorized as follows:

- Benefits resulting from better planning, e.g., from
  - better design of the supply network (including locations, modes of transport, and selection of suppliers),
  - more precise capacity determination of factories, warehouses, and transport systems, and
  - logistics-oriented product and packing design.
- Benefits of integrating operative systems.
  Benefits of support for day-to-day decision making by providing optimization algorithms or heuristics, e.g., for assignment of decisions with respect to available products or capacities (cf. ATP and CTP in Section 4.2.5) or for short-term adjustments of capacities.
- Benefits of better information, e.g., information exchange between supply chain partners or information about events in another entity in the supply chain passed on by IT systems, perhaps with a diagnosis of why this event occurred and what its consequences may be.
- Benefits of systems that audit the compliance with regulations and contracts and may generate early warning signals by applying efficient forecasting techniques, data mining, or filtering mechanisms.

The benefits of SCM systems are difficult to quantify owing to the large numbers of influential factors. Analytical models give some impression about potential benefits in highly simplified environments. As supply chains are often too complex to be studied analytically, several researchers have tried to identify the impact of shared information by using discrete-event simulation models. An overview of results obtained with such models is given by Schmidt and Knolmayer (2006). Unfortunately, the results of the simulation studies do not coincide. A reason for this phenomenon may be that the assumption of a normally distributed demand in upstream companies is violated; causes of this violation are lot sizing and unintended effects of temporal coordination (Schmidt 2007).

In an event study on the impact of SCM systems on corporate performance, Hendricks et al. (2007) recognize that on average adopters of SCM systems experience positive stock market returns, as well as improvements in productivity. Statistically significant improvements are observed in both the implementation and the post-implementation periods.

Consultancies often distinguish between "best-in-class" and average companies. The Performance Measurement Group (2006) claims that companies that are characterized as high SCM performers show the following advantages:

- Service and reliability performance
  - Best-in-class performers have an on-time delivery performance advantage of about 13–25% (the request date) and 8–10% (the commit date) over their peers.
  - While typical companies realize forecast accuracies of 75–80%, best performers exceed 95%.
- Responsiveness and flexibility performance
  - Best performers fulfill their customers' orders 5 times as fast as average companies.
  - Best performers also operate in a more flexible way in supply chains that are able to respond more quickly to unanticipated swings in demand. Their production ramp-up lead times are 8 times as fast as those of average performers.
- Cost and asset performance
  - Best performers consistently have lower operating costs (less than 50% those of average companies).
  - Best performers operate their supply chains with inventory levels that are 65% lower than those of their counterparts. This gives them a significant advantage in the overall performance of working capital, as reflected in the cash-to-cash cycle time.
  - Best performers achieve higher returns on their fixed assets, which drives better shareholder return and revenue growth.

Table 1.1 Improvements in Key Performance Indicators by implementing supply chain functionalities

Company	KPI	KPI value before implementation	KPI value after implementation
Colgate	On-time delivery for Vendor Managed Inventory (VMI)	70%	98%
Colgate	Perfect customer order fulfillment rate	80%	95%
Colgate	Overall order cycle time	9 days	5 days
Coca Cola de Mexico	Forecast accuracy	70%	95%
Coca Cola de Mexico	Truck fleet scheduling accuracy	90%	98%
Hylsa	Forecast accuracy	40%	80%
Cerveceria National	Stock-outs (beer) Stock-outs (soft drinks)	12% 17%	3% 7%
Cerveceria National	Days of inventory (beer) Days of inventory (soft drinks)	7.5 days 10.6 days	5 days 8.3 days

Already for early releases of mySAP SCM remarkable improvements of Key Performance Indicators (KPI) are attributed to the implementation of the system (cf. Table 1.1, based on Chatterjee 2001; Gassmann 2001; SAP 2000).

Goodyear, the world's largest tire company, implemented SAP APO™ to co-ordinate its three SAP R/3™ systems in Europe. Goodyear achieved better transparency within its entire supply chain and reduced

- transport and stock keeping costs by 20–30%,
- order processing times by 25%, and
- erroneously processed orders by 80% (IDS Scheer 2005).

1.3   Risks and Obstacles of Supply Chain Management

The business literature, vendors of SCM systems, and IT consultancies are focusing primarily on the potential advantages of SCM; only a few authors bring up arguments against the prevalent euphoric paradigm (Eßig 2006; Bretzke 2006):

- The SCM literature, trade press, and consultancies are criticized for describing a logistical utopia. Reasons for this viewpoint are:
  - Most collaboration efforts occur between organizational entities that belong to the same group and not between legally independent companies.
  - Typically only two (and not several) entities collaborate intensively.
  - Many companies are part of several (polycentric) supply chains, as otherwise they could not realize economies of scale.
  - Supply chains look different from the viewpoints of the supplier and the customer.
- Close collaboration with supply chain partners may result in inflexibility. The advantages of market mechanisms, the "economics of substitution," and the consideration of aggressive suppliers offering an innovative product spectrum get lost and result in opportunity costs. Can the integration benefits attributed to SCM compensate for waiving these mechanisms? Especially in dynamic environments, the adaptation of plans to accommodate unexpected changes becomes time-intense for organizational (and not primarily IT) reasons. It has been argued that the trend to e-Business and B2B marketplaces makes switching between several business partners easier and that this could lead to an "end of the supply chain" (Singh 1999).
- Egoistic or unethical behavior of participating companies may result in
  - (at least temporarily) pursued goals of a partner that conflict with the objectives of the supply chain,
  - dissemination of biased data,
  - circulation of partners' information to third parties,
  - strategies for profiting from shared data at the expense of the partner supplying this information, and
  - withdrawal from the supply chain at an inappropriate point in time.

- Companies typically behave in a risk-averse manner. Such behavior in a particular supply chain entity may negatively influence results that would otherwise be favorable for the supply chain as a whole. Therefore, Supply Chain Risk Management has to consider means helping companies to take more risk in the interests of the supply chain, by promising and providing support to this company if the high-risk situation should come about. Means include contracts that arrange buy-back procedures, revenue sharing, and cost sharing (Simchi-Levi et al., 2008).
- SCM may highly improve the shareholder value of one company and only marginally improve or even reduce the value of another cooperating company. In the latter case, the company faring less well will only participate in the supply chain if agreements are in place that make the cooperation attractive for every member of the supply chain. However, rules for such a redistribution of values may be difficult to agree upon and to implement. One of the rules would have to compensate opportunity costs resulting for a supplier that could sell its products to third parties at higher margins but has to deliver to the supply chain partner.
- One advantage of close collaboration between partners is seen in the option to reduce time and inventory buffers. However, buffered systems are less vulnerable to structural changes (such as an economic boom, which may result in production and delivery constraints that were inconceivable a few years ago) and to unexpected events. Buffers also allow less extensive exception management, because the number of exceptions to emerge drops.
- Small and medium-sized companies (SME) with inadequate personnel and/ or financial resources could be forced into Supply Chain systems by larger business partners even if their resources are insufficient for this type of collaboration.
- Special properties of some industries may be difficult to represent in industry-neutral software systems for SCM. Software vendors are trying to overcome these problems by service-oriented architectures.
- The implementation of (e.g.) a conventional ERP system is an intra-company project for which "only" units belonging to the same management hierarchy need be coordinated. In contrast, SCM requires that project portfolios, resources, priorities, and plans are coordinated between several, possibly independent, entities. Unwillingness or delays in one company can affect the activities of the partners.
- The development and implementation of SCM software is a major undertaking, with all the risks relating to the observance of design goals, deadlines, and costs.

If SCM obstacles are "solved" with myopic fixes, unintended reactions may result (cf. Table 1.2, based on Lee and Amaral 2002).

In summary, the benefits and risks of SCM have to be critically evaluated. The case studies in Chapter 5 show that in many companies this evaluation results in the decision to improve SCM and to invest in IT systems that support SCM.

Table 1.2 Unintended consequences of myopic fixes of supply chain problems

Example of supply chain problems	Myopic fix	Potential unintended consequence
Late customer shipments	Preferentially expedite "critical" orders	Production disruptions and delays resulting in even more "critical" orders
High material costs	Source from low-price suppliers	Increased scrap and return rates resulting in customer dissatisfaction and high costs
Poor incoming material quality	Hold additional buffer inventory for inbound materials	Higher storage, inspection, and obsolescence costs
Unmanageable proliferation of Stock Keeping Units (SKU)	Increase product commonality	Lower product distinctiveness differentiation leading to lost market share

Chapter 2

The Scope of Supply Chain Management

2.1 Collaboration in Supply Chains

2.1.1 Insufficient Collaboration Results in the Bullwhip Effect

The key feature of SCM is close collaboration between two or more business partners. One of the goals aspired to is to smooth processes and to avoid unpredictable ordering behavior of the main customers; more specifically, to avoid the upstream demand amplification already studied in System Dynamics models (Forrester 1961) and popularized as the bullwhip effect (Lee et al., 1997a, b). The first company to report this phenomenon was Procter&Gamble, which it observed in its diaper supply chain. The most prominent model showing the bullwhip effect is the Beer Game (Sterman 1989). Delays in transferring order information and in fulfillment (due to lead times) and the absence of information sharing are main reasons for the bullwhip effect.

To reduce the bullwhip effect, the members of the supply chain may try to improve their information systems and/or their physical systems. Since the speed of data transfer technology has been dramatically improved in recent years, the assumptions prevalent in the Beer Game about the delays in information transfer can only stem from administrative processes in order management. Data is typically not transferred in real-time, and the coordination effort resulting from the using of different systems may also contribute to time-lags. Furthermore, if the demand is static and normally distributed, there is no reason to order distinct volumes at different time points. If the retailer ordered steadily, the other companies would not have to react nervously to unexpected order volumes. Thus, the bullwhip effect is at least partially homemade.

The main implication of studying the demand amplification is that transferring Point-of-Sales (POS) data to the other partners in the supply chain will considerably reduce the bullwhip effect. However, the question arises why a retailer should share its POS data with other members of the supply chain. One argument is that the supply chain is becoming more competitive, by realizing smoother planning, scheduling, and execution processes. The retailer may also agree to provide the POS data if it assumes that this supportive behavior will result in lower purchase prices or, at least, improve its bargaining power. Furthermore, data about capacity, capacity usage, and inventory may also be shared and be beneficial for the downstream companies. Simulation studies show that the information exchange typically is more important for upstream than for downstream companies (Chatfield et al., 2004).

With respect to collaboration, several maturity levels of supply chains have been defined:

- Stage 1: Functional Focus: Operating discrete supply chain processes with functional management of resources. Supply chain processes and data flows are well documented and understood.
- Stage 2: Internal Integration: Company-wide aligned and integrated supply chain processes continuously measured and steered to achieve common objectives.
- Stage 3: External Integration: Collaboration with strategic partners (customers, suppliers, and service providers) including joint objectives, shared plans, common processes, and performance metrics.
- Stage 4: Cross-Enterprise Collaboration: Information Technology and e-business solutions resulting in real-time planning, decision making, and execution of customer requirements (Roussel and Skov 2007).

The data recorded in the course of the survey shows that only a few companies realize collaboration beyond stage 2; thus, today collaboration between independent legal entities is not very common. However, it should be recognized that the evolution does not necessarily follow this sequence and that some stages (in particular stage 2) may be skipped.

SCM and sourcing decisions are closely related. The number of suppliers may be reduced when a supply chain is designed. In an idealistic view, single sourcing would be appropriate for parts that are offered by supply chain partners. However, risk management may contradict a single sourcing policy. Globalization has a huge impact on achieving supply chain goals. Sometimes offshoring decisions are based on rather myopic views on direct production costs, neglecting such matters as the total cost resulting in the supply chain and the impact on lead times.

2.1.2   Types of Collaboration

2.1.2.1   Information Exchange

Information access and data transfer are highly recommended in SCM systems. Information exchange is bidirectional, while information transfer may be uni-directional. As the company delivering data may not know whether the data transferred or exchanged is relevant for the recipient, the terms data exchange and data transfer would be more suitable. Transfer or exchange of data does not necessarily imply that the recipient is using this data. Therefore, data transfer does not imply that the planning processes of the supply chain partners are based on consistent data. A simplified morphological box distinguishing different types of data exchange is shown in Table 2.1.

Table 2.1 Types of data exchange

Data characteristics	Occurrences			
Source of data	Last element in supply chain (retailer, OEM)	Tier-1 supplier	Tier-2 supplier	…
Recipient of data	Next organization upstream	Next but one/two … organizations upstream	Next organization downstream	Next but one/two … organizations downstream
Category of data	Actual data	Forecast data	Planning data	Meta data
Amount of data	All data	Selected data, defined statically	Rule-based selected data	—
Granularity of data	Elementary data	Aggregated data	—	—
Type of provision	Data access (pull)	Data transfer (push)	—	—
Timeliness	Time-point	Period	—	—
Up-to-dateness	Real-time data	Delayed data, delay time-based	Delayed data, delay rule-based	Delayed data, delay resolved ad hoc

Actual data may be about (e.g.)

- sales volumes at POS,
- inventories,
- warranties,
- capacity usages,
- events, and
- compliance issues.

Planning data concern (e.g.)

- strategies,
- investments in physical systems and information systems,
- events such as promotions, announcements of end-of-life products, or of new product introduction,
- procurement,
- production,
- scheduling,
- distribution, and
- financial matters.

Meta data may be exchanged to coordinate

- quality control, and
- the use of IS, in particular the
  - customization of IS,
  - data models,
  - process models, and
  - numbering systems.

Another type of data transfer tries to improve the capabilities of the suppliers, for example with respect to product quality.

Mini case: Nestlé supports sustainability in the supply of agricultural raw materials and agricultural best practices. To translate its words into actions, Nestlé employs over 800 agronomists, technical advisers, and field technicians. Their job is to provide technical assistance to more than 400,000 farmers throughout the world to improve their production quality, as well as their output and efficiency. They do this on a daily basis in as many as 40 countries. This specialist team has pioneered the development of sustainable local fresh milk and coffee production (Nestlé 2006).

2.1.2.2 Collaborative Forecasting

Collaborative forecasting is based on data exchange or transfer, but does not necessarily result in collaborative planning. This distinction is also emphasized in the CPFR model (cf. Section 2.1.2.4). The goal of collaborative forecasting is to find a consensus on future data that may be used in local planning or in collaborative planning efforts.

The Delphi method is a well-known procedure for collaborative forecasting of future trends. Results show that divergent opinions of experts converge some way toward a consensus when those involved are informed about opinions expressed by other experts. However, the result of applying the Delphi method is not a forecast accepted by all concerned. The Delphi method is typically not used in routine forecasting of operative data but in forecasting future trends. Application of the Delphi method can be supported by specific IT systems.

Achieving a common forecast of quantitative data, for example about future demand for certain products or product groups, is a difficult task. Planning typically means considering distinct scenarios that differ in the assumptions and data underlying them. A company may look at several scenarios, and the common forecast may be just one of several considered. An agreement to use only a consensus forecast may reduce the value of local planning processes considerably and cannot be enforced.

2.1.2.3 Collaborative Planning

Collaborative Planning aims to coordinate the plans of several partners in the supply chain. The associated models can be managed by one or more of the firms involved or by a trusted service provider.

Several types of models may be used. Spreadsheet models and simulation models may be developed to show the consequences of different decisions in certain planning scenarios ("What-If Models"). How-to-Achieve Models change the perspective by stipulating target values and determining the corresponding value of an independent variable. Decision models are used to determine the best solution by optimizing algorithms or to find a satisfactory solution by applying heuristics.

Collaborative planning differs from individual planning in several ways (Table 2.2, partially based on Windischer and Grote 2003).

Table 2.2 Comparison of individual planning and collaborative planning

Individual planning	Collaborative planning
Recognizing the sequential order of events	Communication of anticipated events
Recognizing goals	Lateral agreements on goals
Recognizing the availability of alternatives	Information exchange about the availability of alternatives
Recognizing the adequacy of plan's resolving	Recognizing the adequacy of common plans
Monitoring planned actions and diagnosing errors in individual plans	Monitoring and diagnosing errors in common plans
Revising individual plans	Coordination of planning and feedback about modifications
Canceling individual plans	Common reflection and common decisions to cancel plans

Depending on the amount of information transparency agreed upon, several types of collaborative planning can be distinguished. One of them is Open Book Planning. The collaborating entities deliver data into a common planning model. The semantics of this data (i.e., the definitions used in the data models) must be carefully coordinated. The data and the results obtained by the planning procedure become visible to all participating entities. A very high level of trust is necessary between the partners for this approach to be realized. Even entities belonging to the same group may have objections against (detailed) Open Book Planning. The Open Book may be accessible only to selected members of the supply chain. However, in such a situation it may be even more difficult to make sure that the other entities deliver correct planning data.

Another approach is to install a trusted service-provider as the entity collecting data for the planning model and delivering the planning results to the supply chain partners. In this case the cooperating entities are treated equally with respect to information transparency. However, results of a planning model are usually not implemented without further consideration. The purpose of decision models is to provide insight, not numbers. Insight is based on understanding relationships between input data and output data. It may be difficult to gain insight if the effects of modifying input data cannot be discussed in detail because the input data is clandestine.

A common planning model may become complex owing to its size and the details considered, and it may be difficult to find appropriate algorithms for determining an optimal solution or even for applying a sound heuristic. Decomposition has been recommended to reduce the complexity of decision models. In this case it is not necessary to exchange all details of the data relevant to the planning model, but only some results obtained from local planning models.

Decomposed decision models are solved in an iterative way. The results of the planning model P_{ir} of entity i in iteration r are used by the collaborating entity j in iteration $r + 1$. Entity j will consider the effects of P_{ir} on its own situation and decision variables and develop plan $P_{j, r+1}$, which is communicated to entity i. Thus, the planning results of one entity appear as input data in the plan of the other entity.

For obvious reasons only a limited number of such organizational iterations can be realized. The optimal solution, which could be determined by an Open Book model, will typically be missed. However, numerical experiments show that even a small number of organizational iterations may result in solutions that are quite close to the optimum of the Open Book model and, from the perspective of the supply chain, far better than local solutions obtained without collaborative planning (Dudek 2004; Dudek and Stadtler 2005).

2.1.2.4 Collaborative Planning, Forecasting, and Replenishment (CPFR)

Several frameworks for structuring collaboration tasks exist. The best known is the CPFR® framework. CPFR® is a reference model developed by the Voluntary Interindustry Commerce Solutions (VICS) Association. Fig. 2.1 shows that the CPFR model distinguishes eight collaboration tasks. For collaboration between a retailer and a manufacturer the tasks are exemplified in Table 2.3.

2.1.2.5 Collaborative Scheduling

As scheduling decisions are often short term and taken close to execution, real-time information exchange and contingency management among geographically dispersed entities may be beneficial (Jia et al., 2002; Boyson et al., 2003).

The schedule of transports may determine production schedules, and a need for the exchanging of information between distribution and production schedulers results (Chen and Vairaktarakis 2005). The customer may receive information about successfully finished operations and the time intervals for which remaining operations are scheduled. This could be done via alerting mechanisms (e.g., sending e-mails or messages to a PDA), by providing information on the Web, or even by allowing access to (parts of) the partner's scheduling system.

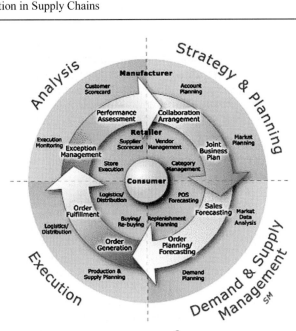

Fig. 2.1 Visualization of the CPFR® process (VICS 2004)

Table 2.3 Collaboration tasks between a retailer and a manufacturer (cf. VICS 2004)

Manufacturer Tasks	Collaboration Tasks	Retailer Tasks
	Strategy and Planning	
Account Planning	Collaboration Arrangement	Vendor Management
Market Planning	Joint Business Plan	Category Management
	Demand and Supply Management	
Market Data Analysis	Sales Forecasting	POS Forecasting
Demand Planning	Order Planning/Forecasting	Replenishment Planning
	Execution	
Production and Supply Planning	Order Generation	Buying/Re-buying
Logistics/Distribution	Order Fulfillment	Logistics/Distribution
	Analysis	
Execution Monitoring	Exception Management	Store Execution
Customer Scorecard	Performance Assessment	Supplier Scorecard

Mini case: In the chemical industry, changes in the schedule of one plant can affect several other plants, and ripple effects may increase the magnitude of changes in plants downstream. For instance, in the Bayer company the plant schedules are highly interdependent. The results of the nightly centralized scheduling run are broken-down into plant-specific models where decentralized planners use these models for local changes. The local scheduling persons should

- *be able to work on a smaller model of the facilities they are allowed to schedule but at the same time be able to share data with and view information from other plants,*
- *be able to see the schedule changes of relevant production steps in other plants,*
- *make other plants aware of schedule changes, and*
- *reduce conflicts and find a mutually agreeable solution for product chains running through multiple plants with the help of a chain planner.*

Complex communication mechanisms are set up to achieve these goals. Central coordination mechanisms are combined with complementary information exchange amongst decentralized decision makers between the scheduling runs (Berning et al., 2002).

2.1.2.6 Collaborative Execution

Collaborative execution may be closely connected with reassignment of tasks and resources and the redesign of physical processes. In this case not only information and planning systems are influenced by SCM but also the physical execution systems.

Changes of physical systems have been suggested by such production management concepts as Just-in-Time (JIT), Lean Production, and Agile Manufacturing. JIT needs close collaboration between the partners, and reducing setup times is an important precondition for the realization of JIT procedures. Cross-docking is a concept intended to minimize handling times at distribution centers by tight co-ordination of inbound and outbound transports. Track&Trace systems show the progress made in bridging the spatial distance between supplier and recipient and allow the recipient to prepare for arrivals, but also to adjust production schedules if an item required should arrive too late. Visibility of real-time data for business partners is regarded as one of the main properties of a "real-time enterprise." Many SCM systems support the visualization of data.

Mini case: Several companies with basically decentralized organizational structures achieved significant improvements through central coordination of material handling. For instance, the largest Swiss retail company Migros helped to develop an Application Service Providing (ASP) platform for achieving better visibility and transparency of the associated pallet flows between its suppliers, the central regional warehouses, and its supermarkets (Knolmayer and Dedopoulos 2006).

2.1.2.7   Collaborative Monitoring and Controlling

Mini case: In the 1980s, General Motors' Service Parts Operation used sophisticated Operations Research methods for inventory and transportation management in its relationships with dealers. However, the service to consumers was consistently poorer than the service of most of its competitors, because the dealers' inventory systems were out of control, resulting in outdated data and metrics and wrong stock-keeping decisions. This illustrates the fact that a supply chain is only as good as its weakest link (Hausman 2004).

Many criteria have been proposed for measuring and evaluating the performance of a company's logistical system. Examples are

- (differently defined) service levels,
- response delay, the difference between the delivery day initially requested by the customer and the negotiated day,
- lateness, computed from the differences between negotiated delivery day and actual delivery day,
- (differently defined) stocks, e.g., work in progress (WIP) as a percentage of sales,
- mean and variance of throughput times, and
- percentages of scrap in production and corrupted inventory.

The Supply Chain Operations Reference (SCOR®) model developed by the Supply-Chain Council (SCC) defines more than 200 Key Performance Metrics at the highest of four levels. SAP SCM™ provides more than 300 KPI that are based on the SCOR® metrics. Three classes of customer-facing and two classes of internally oriented performance attributes are distinguished (cf. Supply Chain Process Improvement 2007):

- Customer-facing performance attributes
  - Reliability
    - Delivery performance
    - Perfect order fulfillment
    - Fill rates
  - Responsiveness (Order fulfillment lead times)
  - Flexibility
    - Supply chain response time
    - Production flexibility
- Internal-facing
  - Costs
    - Costs of goods sold
    - Total SCM costs
    - Warranty/returns processing costs
  - Asset management efficiency
    - Cash-to-Cash cycle time
    - Asset turn.

The Supply Chain Performance Indicator, which has been defined by The Performance Measurement Group (2007), considers a broad spectrum of business-related metrics which shows the high impact of good SCM practices on business results. With respect to the large number of metrics it is recommended that the most relevant ones be selected. These may be visualized on a dashboard, using Kiviat graphs, spider diagrams, or Balanced Scorecards (Kaplan and Norton 1996).

In addition to the SCOR® model, a Design Chain Operations Reference (DCOR) and a Customer Chain Operations Reference (CCOR) model have been defined by the SCC. These models also define many metrics. A projection of some metrics to Balanced Scorecard Categories is suggested by Bolstorff (2006). Ways of projecting the SCM metrics into terms of income statements, balance sheets, and Economic Value Added indicators have also been suggested (Camerinelli and Cantu´ 2006).

For supply chains, two different controlling approaches exist. On the one hand, each entity in the supply chain can define its own criteria and eventually present the values achieved in a Balanced Scorecard; however, if this information is passed on to partners, a shared meaning should be accomplished, and this can only be reached when there is agreement upon the definition of data elements and co-ordinated procedures are applied. On the other hand, common metrics for the whole supply chain may be defined and eventually presented in a Supply Chain Scorecard; coordination of meta data becomes even more relevant when this approach is followed (cf. Ackermann 2003; Kleijnen and Smits 2003).

2.1.2.8 Collaborative Reassignment of Tasks

The most far-reaching type of collaboration is a coordinated restructuring of functions and processes, which may result in reassignment of task responsibilities from one supply chain partner to another. In redesigning a supply chain, inter-mediation or disintermediation may be considered when tasks are reallocated; one example of such an approach is the Fourth-party Logistics Provider (4PL) concept. Quality control can be moved from the customer to the supplier after a common quality management system has been agreed on. Financial flows can be reorganized by applying Electronic Bill Presentment and Payment (EBPP) systems (SAP 2001) as part of Financial Supply Chain Management.

Vendor Managed Inventory (VMI) is probably the most common reassignment of responsibilities. The customer is no longer placing orders and, therefore, no due dates for delivery are fixed. The vendor is responsible for providing concerted inventory service levels. SAP recommends considering VMI if

- key customers constitute a high percentage of the vendor's sales figures,
- the products are standardized and requested repeatedly,
- product growth is not excessive, meaning that the requirement patterns are stable and the vendor can assume that requirements will not occur spontaneously, and
- the transaction costs for order processing and production planning are high (SAP 2007).

Intentia (2001), a former vendor of ERP systems, describes benefits of VMI as follows:

- Supplier benefits
  - Visibility of the customer's POS data simplifies forecasting.
  - Promotions can be more easily incorporated into the inventory plan.
  - Customer ordering errors, which in the past would often lead to a return, are reduced.
  - Stock level visibility helps identify priorities (replenish stock versus a stockout).
  - The supplier can see the potential need for an item before the item is ordered.
- Customer benefits
  - Fill rates from the supplier, and to the end consumer, improve.
  - Stockouts and inventory levels often decrease.
  - Planning and ordering costs decrease since the responsibility is shifted to the supplier.
  - The overall service level is improved by having the right product at the right time.
  - The supplier is more focused than ever on providing superior service.
- Dual benefits
  - Data entry errors are reduced owing to computer-to-computer communications.
  - Overall processing speed is improved.
  - Both parties strive to offer better service to the end consumer. All parties involved benefit when the correct item is in stock when the end consumer needs it.
  - A true collaborative partnership is formed between the supplier and the customer.

Extremely high benefits are reported from realizing VMI relationships. SAP claims very optimistic figures that have been reported by SAP customers or independent third-parties (Table 2.4, cf. SAP 2007).

Mini case: Knorr-Bremse, a leading producer of brake systems with more than 60 locations in 25 countries, implemented the SAP Inventory Collaboration Hub™ in 2005. The costs of order processes and administration expense were reduced by more than 50%. Many A and B materials are stored via Supplier Managed Inventory agreements. Capital lock-up was reduced by lower warehouse inventory and safety stocks (Brauchle 2006).

Table 2.4 Potential benefits of VMI (SAP 2007)

Business benefits	Vendor/customer	Value potential
Increased revenue/sales	Vendor and customer	100–200%
Lower inventory levels	Vendor	70%
Increased service levels	Vendor	From 89% to 98%
Operating costs through full truckloads	Vendor	28%
Increased service levels	Customer	From 93% to 99%
Inventory turns	Customer	27–70%
Increased service levels	Customer	From 93% to 99%

When considering the potential of VMI one has to realize that it is based on the transfer of detailed data, e.g., POS data and inventory data. Such data transfer may be realized with or without entering on a VMI relationship. An advantage of VMI is that no due dates are fixed by the customer, which makes the vendor flexible with respect to its execution processes. However, the vendor may lack some information which is available only locally at the site of its customer. VMI partnerships should incorporate the obligation to transfer either such local information or at least forecast data based on it. Several simulation studies on VMI systems show significant cost reductions for the entire supply chain (Disney and Towill 2003a, b). As suppliers have access to actual sales and/or inventory data provided by the customers, the Bullwhip Effect can be reduced, resulting in a smaller variability of demand data (Småros et al., 2003). Thus, safety stocks, particularly of suppliers, can be reduced.

Sometimes a distinction is made between Vendor Managed Inventory and Supplier Managed Inventory (SMI). In the CPFR® context four alternatives are distinguished (Table 2.5). The difference between VMI and SMI is primarily one of viewpoint: VMI involves the management of finished goods inventories outbound from a manufacturer, distributor, or reseller to a retailer, whereas SMI manages the flow of raw materials and component parts inbound to a manufacturing process (Pohlen and Goldsby 2003). IT ownership and IT architectures differ. In the SAP environment there is also a difference in the ownership of the collaborative application system – for VMI the application system is owned by the supplier and for SMI, by the customer.

Table 2.5 Assignment of responsibilities (cf. VICS 2004)

Alternative	Sales forecasting	Order planning	Order generation
Conventional approach	Retailer	Retailer	Retailer
Co-managed inventory	Retailer	Retailer	Manufacturer
Supplier managed inventory	Retailer	Manufacturer	Manufacturer
Vendor managed inventory	Manufacturer	Manufacturer	Manufacturer

2.2 Business Architectures for Supply Chain Management

2.2.1 Supply Chain Planning Matrices

Several models have been followed to arrange the most relevant SCM processes in systematic frameworks. In the research literature several (slightly different versions of) Supply Chain Planning Matrices (Fig. 2.2) are presented (cf. Neumann et al., 2002; Fleischmann and Meyr 2003; Fleischmann et al., 2005; Meyr et al., 2005).

A detailed description of the matrix is given by Fleischmann et al. (2005, p. 88). In our opinion, Supply Chain Planning Matrices have some disadvantages. The arrows in the top row imply that a certain flow occurs independently of the type of production system. However, for make-to-order production the sequence of the columns "Production," "Sales," and "Distribution" should be "Sales," "Production," and "Distribution," and order-specific design activities for make-to-engineer production should appear. Furthermore, the execution and controlling processes are disregarded in the framework and the collaboration with other companies is not visualized in the matrix. We try to improve these shortcomings in our pyramidal representation (cf. Section 2.2.3).

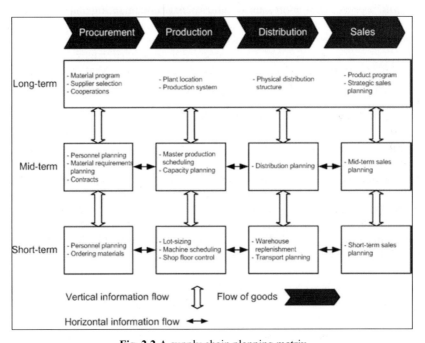

Fig. 2.2 A supply chain planning matrix

2.2.2 *The Supply Chain Operations Reference Model*

In practice, the Supply Chain Operations Reference (SCOR®) Model developed by
the Supply-Chain Council is attracting a lot of attention. Today the Council counts
about 1,000 corporate members worldwide and has established chapters in North
America, Europe, Greater China, Japan, Australia/New Zealand, South East Asia,
Brazil, and Southern Africa. In 2008, Release 9.0 of the SCOR® model was made
public (Supply-Chain Council 2008). SAP AG is a member of the SCC and a main
sponsor of its activities.

The SCOR® model is a process reference model, proposing a certain terminology
and system of notation for describing business processes. It is organized into four
levels to allow differently detailed views on business processes and focuses on
inter-organizational processes. A company or a supply chain may use the SCOR®
model to describe the current status of the system ("as-is" situation) or to define a
target status ("to-be" situation). Such models are often used in business process
reengineering projects. The SCOR® model also defines metrics used to measure the
performance of certain process elements. A company may decide to gather this
data for internal performance evaluation or also for benchmarking with other
companies. The SCC tries to motivate its members to deliver performance data for
the SCOR® metrics to support inter-organizational benchmarking and to recognize
"best practices."

The SCOR® model defines five process types

- Plan
- Source
- Make
- Deliver
- Return

at four hierarchical levels. At the uppermost level, the process types are defined as
shown in Table 2.6. According to the SCC, level 1 of the SCOR® model aims to
support companies in making basic strategic decisions regarding its operations in
the following, sometimes vaguely formulated areas:

1. Delivery performance,
2. Order fulfillment performance,
3. Fill rate (make-to-stock),
4. Order fulfillment lead time,
5. Perfect order fulfillment,
6. Supply chain response time,
7. Production flexibility,
8. Total SCM cost,
9. Value-added productivity,
10. Warranty cost or returns processing cost,

11. Cash-to-cash cycle time,
12. Inventory days of supply, and
13. Asset turns.

At level 2, e.g., the Make process is refined to

- Make-to-stock production,
- Make-to-order production, and
- Make-to-engineer production,

whereas the Return process is detailed to

- Return defective product,
- Return Maintenance, Repair, and Overhaul (MRO) product, and
- Return excess product.

Level 2 also defines some enabling processes. A typical example of an enabling process is to provide the necessary IT infrastructure for process execution.

In 2007, the SCC announced SCORmark^{SM}, a members-only benchmarking portal based on the SCOR$^®$ model, in association with APQC (Supply-Chain Council 2007). As part of the "analyze" phase of the SCOR$^®$ model, a company may use SCORmark^{SM} benchmarking

- to select the supply chain metrics most critical to its organization,
- to determine the target performance desired for each supply chain attribute in the SCOR$^®$ model, and
- to enter the relevant data required to calculate the performance for each selected metric into the secure, confidential benchmarking portal.

Table 2.6 Level 1 Processes, as defined by the Supply-Chain Council

SCOR$^®$ process	Definition
Plan	Processes that balance aggregate demand and supply to develop a course of action which best meets sourcing, production, and delivery requirements.
Source	Processes that procure goods and services to meet planned or actual demand.
Make	Processes that transform products to a finished state to meet planned or actual demand.
Deliver	Processes that provide finished goods and services to meet planned or actual demand, typically including order management, transportation management, and distribution management.
Return	Processes associated with returning or receiving returned products for any reason. These processes extend into post-delivery customer support.

The data is validated in a seven-step process to produce a report with

- an executive scorecard to quickly spotlight on any gaps in the targeted performance levels for each supply chain attribute and
- a detailed analysis for each specific metric selected, including best practice information on the drivers of performance and peer group reporting as available.

ARIS™ is a business process management tool developed by IDS Scheer and today offered as part of SAP's NetWeaver infrastructure. Among other tools, an ARIS EasySCOR Modeler has been developed (IDS Scheer 2007).

Software vendors included SCOR® metrics in their SCM systems (Gassmann 2001). The SAP Solution Manager is an implementation tool that allows mapping of the SCOR® model's "best practices" against what the users want in their SAP ERP™ and SAP SCM™ implementations. Once in operation, the SAP ERP™ and SAP SCM™ systems automatically deposit the data from ERP and ongoing supply chain transactions into SAP's Business Intelligence™ applications. These calculate the plan-source-make-deliver KPIs and deliver them to SAP's management cockpit for role-based breakdowns of the SCOR® model (Gould 2005).

Some deficiencies of the SCOR® model as seen from an academic point of view are discussed by Huan et al. (2004) and Poluha (2007).

2.2.3   A Supply Chain Pyramid

Based on the Supply Chain Matrices and the SCOR® model, we present a global view of the supply chain tasks in the form of a pyramid (Fig. 2.3) and use this pyramid as a reference framework. With the pyramidal form we reflect the hierarchy of decision rights, planning tasks, and the associated information needs (as often visualized in "information pyramids"; cf. Mertens 2007, p. 6). A slightly similar "task reference model of transcorporate logistics" in pyramidal form was proposed by Hieber (2002).

Fig. 2.3 shows inbound- and outbound-collaboration tasks at various organizational levels. Strategic, tactical, and operational planning tasks are distinguished at the horizontal levels. For ease of presentation, operational planning and scheduling are combined at one level. Source and procurement and make and production are used synonymously. Execution is explicitly included in the pyramid.

Compared with the SCOR® model, product design is enclosed in the pyramid to avoid the formulation of a separate design model. Furthermore, the SCOR® model does not explicity address sales activities (which later became part of the widely neglected CCOR). To emphasize the high importance of selling, we decided to split the

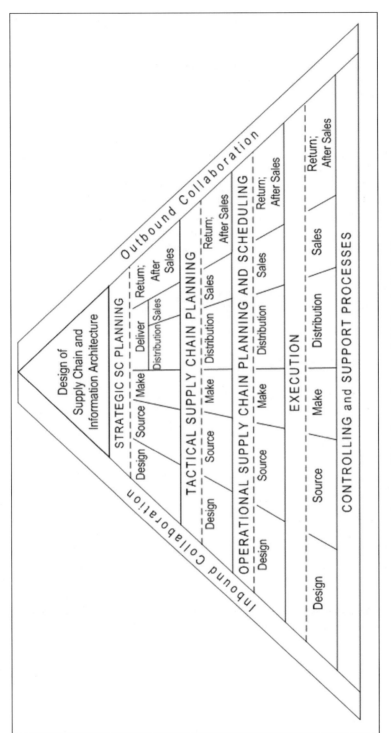

Fig. 2.3 Pyramidal reference framework for Supply Chain Management

delivery process into distribution and sales already in our Supply Chain Pyramid. We avoid arrows to indicate that sequences between sales and distribution activities depend on the type of business relationships. Finally, we think that regular after-sales processes are at least as important as return processes and therefore decided to mention them explicitly in the SCM pyramid. These modifications are in line with the Supply Chain Visibility Model proposed by the IBM Institute of Business Value (Butner 2007).

The pyramid is refined stepwise in the remainder of the book. First, SAP's SCM solution map is projected onto the SCM pyramid (Fig. 3.4). As we shall see, tactical and operational planning are well covered by the SAP system, as well as controlling and support processes. Some strategic planning tasks are difficult to support by means of IT systems, and there are some uncovered spots on the landscape. Product design activities are outside the scope of SAP's systems, but there are interfaces to the most relevant CAD systems.

In Section 2.3 we formulate desirable properties of SCM systems. In Chapters 3 and 4 we discuss how these properties are covered by the SAP SCM™ 5.0 system.

2.3   Desirable Features of SCM Systems

This section gives a short description of some functionalities that could be helpful for SCM. We do not describe functions that are also essential for enterprises not involved in supply chains.

Functions and processes in SCM systems have specific characteristics because

- data of several networked enterprises, not only those of one company, must be stored. There is a need for special filtering and compression mechanisms before data is fed into common databases, to avoid too great an increase in the sizes of databases and data warehouses;
- single bodies of data often have to be aggregated into larger groups: for instance, equipment into capacity groups, product characteristics into characteristics groups, products into product groups;
- bigger problems have to be decomposed before they can be treated with optimization algorithms or heuristics; examples are the segmentation of a long planning horizon into several shorter time segments, for which solutions may successively be found although typically the global optimum will be missed.

2.3.1 Design of Supply Chain and Information Architecture

Desirable Features	Comments	Coverage by SAP SCM™
a) Visualization of the complete supply chain/network at several levels of detail.	Powerful tools for visualization and navigation (cockpit, leitstand) are required.	Cf. Section 4.6.2.A.
b) Modifications of the network, e.g., regarding intermediation and disintermediation or new allocation of tasks to the collaborating firms.	When structuring a supply chain, decisions have to be made about which tasks should be carried out by the supply chain partners themselves and which are to be outsourced. With the impact of the Web on the firms' cooperation, new intermediaries (e.g., Trust Centers to support financial transactions) are founded, whereas some traditional intermediaries (e.g., travel agencies) are declining in importance.	For master data relevant for modeling a given supply network see Section 4.7.3.

SAP SCM™ does not provide functionality for decision support in supply network design. |
| c) Configurator to formulate agreements and contracts by combining text modules (e.g., subcontracting, service level agreements, periods of notice, allowance of delays, responsibility for consignment stocks). | When the structure of the supply chain remains fixed for a longer time span during which some partners are replaced, the contracts with new partners can be derived from standardized agreements by modifying certain parameters. Examples are contracts between suppliers of standardized products and big retailers. | Not in scope of SAP SCM™. |
| d) Collaborative investment planning; negotiation of investment proportions. | In supply chains the partners sometimes agree to share large investments, such as warehouses near airports or projects to improve data quality, by using RFID techniques. | Not in scope of SAP SCM™. |

Desirable Features	Comments	Coverage by SAP SCM™
e) Decisions and actions to avoid multiple quality checks.	Multiple quality checks (at the supplier's site before shipping and at the customer's site after receipt) can be avoided if the partners agree on a Quality Management system: Results of quality control measures may be exchanged via a portal where the customer can see them or have them analyzed by an automated system.	Not in scope of SAP SCM™.
f) Navigation details of the master files by activating nodes and arcs such as information on capacity and costs of factories, warehouses, transportation routes, production programs of suppliers, time zones, and holidays at the locations of partners.	Top-down navigation, opening windows, e.g., to visualize such additional information as capacity constraints of suppliers or customs regulations.	Not in scope of SAP SCM™.
g) Management of variants, especially in the context of multiple sourcing.	The decision on how many and what kinds of variants are allowed is difficult because of the impact on sales and the costs of production, inventory, logistics, training, after-sales service, and capital requirements. ABC analyses are recommended for support of variants management. When parts are procured from different suppliers, these parts may not be strictly identical. This may be one reason for variants.	Not in scope of SAP SCM™.

Desirable Features	Comments	Coverage by SAP SCM™
h) Demonstrating the differences between objects [products, product features ("product interchangeability"), production processes] that can be substituted by each other and explaining the implications, e.g., additional costs caused by a substitution.	Substitution processes may be induced by changes in the cost of objects or result from sourcing or production bottlenecks.	Substitution is considered in planning; cf. Sections 4.4.6 and 4.7.3.
i) Positioning of inventory.	Decisions about storage locations of raw materials, components, and products are fundamental in SCM. Criteria include customer demand, service level agreements, transportation time and costs of transportation between sites, value of the parts, the necessary storage space (depending on the volume per item), and special technical conditions, such as cooling or security.	Decisions on whether or not to stock an existing warehouse cf. Section 4.4.2.A. Further analyses are not in scope of SAP SCM™.
j) Agreements on lot sizes and production cycles.	Set-up times and set-up costs may differ considerably between supply chain partners. The situation where one partner produces big lots a few times per year while the downstream partner manufactures small lots more often should be avoided, because such uncoordinated decisions result in avoidable inventories.	Not in scope of SAP SCM™.
k) Management of hierarchies of the sources, including service stations.	*Example: Supplier -> central warehouse -> regional warehouse -> service stations.*	Maintenance as master data; cf. Section 4.7.3. Decision support for supply chain design is not in scope of SAP SCM™.

Desirable Features	Comments	Coverage by SAP SCM™
l) Coordination of data models in databases, data warehouses, and data marts of supply chain partners.	Harmonized data models help to avoid errors in production and comparison of management information, e.g., values of KPI.	Not in scope of SAP SCM™.
m) Coordination of process and workflow models.	Coordinated workflow models avoid redundancies, support temporal coordination, and increase transparency in the supply chain.	Alerts can be exchanged; cf. Section 4.6.2.
n) Coordination of customizing parameters.	Determining the parameters of ERP systems and other software packages is a difficult task because of many complicated effects, side effects, and interdependencies. *Example: Minor modifications of the parameters of the function "consumption of plans" for final products may have far-reaching effects on stocks of components and raw materials.*	Not in scope of SAP SCM™.
o) Manifold simulations to evaluate alternatives.	Simulations could support tasks mentioned in items g), i), j), and k) in particular.	Cf. Section 4.2.

2.3.2 Collaborative Product Design ("Design for SCM")

Desirable Features	Comments	Coverage by SAP SCM™
a) Cooperation of the supply chain partners in product design, value analysis, and value engineering.	An SCM-oriented design considers the effect of design decisions on all partners in the supply chain. These effects may not be obvious to the designers of a certain company; therefore, collaboration between design and manufacturing experts of the participating companies will facilitate a "Design for SCM." During the design process several partners should be able to see the actual design results on their screens and to add hints and suggestions. The evaluation of alternative designs by the partners in the supply chain will often differ. Tools, e.g., groupware, for better coordination between supply chain partners may be helpful.	Not in scope of SAP SCM™.
b) Process planning taking the effects on SCM into account.	Examples are the implementation of postponement concepts or requirements of RFID, such as avoiding the interruption of RFID communication by metallic packing or special protection needs of products (e.g., avoidance of damage through sharp profiles during transport).	Not in scope of SAP SCM™.
c) Measures to allow simple transfer and exchange of product-defining data by coordinated application of CAD, PDM, and PLM systems.	If incompatible CAD tools are used, some physical components may not fit together. *Example: Delays in the development of the Airbus Super-jumbo A380 are attributed to the use of different releases of the CAD system CATIA. Interoperability problems resulted from using version 4 in German and Spanish plants whereas factories in France and the UK used version 5 (Steinke 2006).*	Not in scope of SAP SCM™.

Desirable Features	Comments	Coverage by SAP SCM™
d) Manifold simulations to evaluate product and process alternatives.	Modern CAD systems allow the evaluation of some properties of physical products and processes without the necessity for building a physical prototype. *Example: A subassembly designed by the supplier has to be adjusted to suit the tools available in the assembly line of the customer's plant.*	Not in scope of SAP SCM™.

2.3.3 Sales and Demand Planning

2.3.3.1 Strategic and Tactical Demand Planning

Desirable Features	Comments	Coverage by SAP SCM™
a) Portals for collaborative planning (planning book).	The planning book is the main screen where the common data is displayed, entered, and processed and where interactive planning takes place.	Cf. Section 4.2.3.A.
b) Methods of calculating customer lifetime values.	Estimating the customer lifetime value helps in decisions on whether a customer should be invited to become member of a supply chain.	Not in scope of SAP SCM™.
c) Consideration of the product life cycle.	The life cycle profiles have to be adapted to take account of recent developments. The partners should provide relevant data for determining typical profiles for SCM purposes by using statistical methods, e.g., forecasting methods based on market saturation and sales data aggregated over the lifetime of the product.	Cf. Sections 4.2.1 and 4.4.1.C.

Desirable Features	Comments	Coverage by SAP SCM™
d) Coordinated inventory planning for products near to the end of their life cycle.	The exchange of product life cycle data is of utmost importance when it comes to inventory management of products that will soon be eliminated from the sales program. Careful coordination helps to keep the right inventory in the right warehouses, shops, and service stations.	Not in scope of SAP SCM™.
e) Supporting Collaborative Planning, Forecasting, and Replenishment (CPFR).	CPFR is a thorough coordination of functions and processes in the supply chain with potential for considerable benefits to the partners. *Example: Volvo, a Swedish car manufacturer, implemented collaborative processes with over 350 domestic and overseas vendors and suppliers. One of the main functions is the collaborative exchange of forecasts.*	Cf. Section 4.2.3.
f) Methods of VMI or SMI.	The supplier controls the inventory of the customer and replenishes it when necessary. *Example: Woolworth calls up the sales of all items in each of its branches every day after closing time. L'Oréal, the global market leader in the cosmetics industry, determines delivery quantities and creates an optimal delivery plan using the SAP SCM™ supply network planning capabilities based on forecast demand, anticipated stock movements, open orders, and inventory information.*	Cf. Sections 4.2.3.B, 4.3.1.D, and 4.3.2.

Desirable Features	Comments	Coverage by SAP SCM™
g) Collaborative forecasting supported by a method bank with forecasting algorithms and systems to support the selection of methods, parameter configuration, and the interpretation of results.	The partners may use different forecasting procedures, which, moreover, may be modified depending on deviations between forecast, demand, and sales. If collaborative forecasting is practiced within the supply chain, the partners should be able to analyze the details, including the parameter selections and the results, by using an explanation component.	Cf. Section 4.2.1.A.
h) Administration of time series and time series patterns, such as sales after promotions, e.g., separated for customer types and regions ("global forecasting profiles") or patterns of cannibalization.	Often promotions are limited to a region, e.g., dependent on the local weather or local holidays. Promoted products may cannibalize others. The demand may be shifted to later periods because customers build up stocks of the promoted parts and buy less in the following periods. For this reason logistic managers do not favor promotions, but the arguments of the marketing specialists often dominate, so that the problems have to be solved by SCM.	Administration of time series is a basic capability of Demand Planning and Forecasting; cf. Sections 4.2.1 and 4.4.1. Promotion planning is discussed in Sections 4.2.1.D and 4.3.2.C.
i) Incorporating external data.	Data from external databases, e.g., business cycle indicators, should be merged with internal data if this might improve the accuracy of the forecasts.	Incorporation of external data is a basic concept of SAP SCM™ via SAP BI™. Application of external data for forecasting is described in Section 4.2.1.A.
j) Standardized analyses of the accuracy of the forecasts and of the replenishment policy (Forecasting and replenishment analytics).	Common parameters of the analysis methods allow for better diagnosis of appropriate inventory levels in the partner companies.	No standard reports. For stocking decisions cf. Section 4.4.2.A.

Desirable Features	Comments	Coverage by SAP SCM™
k) Variable aggregation of resources such as workers (described by skill codes), materials, production facilities, and transport vehicles. Aggregation of periods, regions, products and product characteristics ("characteristic-dependent forecasting"), aggregation to virtual products ("phantoms"), block planning with buckets. Disaggregation of aggregate planning data, e.g., of total production quantities to countries and factories.	The systems should be able to avoid information overload by providing powerful aggregation procedures. Supply chain partners use different methods and parameters to aggregate, e.g., in order to determine data about product groups with an adequate statistical basis. In different countries the product characteristics with most impact on the market may vary. *Example: It should be possible to cluster all car motors with defined values of the CO_2 output or all toys made from the same raw material.*	Cf. Sections 4.2.1, 4.2.2.C (regarding time), and 4.2.2.N (regarding resources).
l) Forecast after constraints.	If forecasts indicate a significant growth in demand, information on serious capacity constraints should prevent planning with these sales forecasts.	Cf. Section 4.2.1.J.
m) Collaborative demand planning.	The partners in the supply chain should try to develop common scenarios and to achieve a common estimate of the demand resulting in use of these scenarios as the basis of subsequent planning.	Cf. Section 4.2.3.A.
n) Product mix planning.	Suppliers often have an interest in receiving orders not only for single products but for product mixes. *Example: A chemical company is interested in selling product A together with product A', which emerges as a joint product in the production process.*	Cf. Section 4.7.3.

Desirable Features	Comments	Coverage by SAP SCM™
o) Collaborative delivery schedules.	Information exchange before confirmation of shipment dates can increase the probability of shipping in time.	Cf. Sections 4.2.3.C and 4.2.3.D.
p) Collaborative promotion planning.	Promotions may be triggered by inventory management (high inventory levels have to be reduced, additional warehouse space is needed), from special sales opportunities (e.g., TV and video sets before big sporting events), from finance (need to improve liquidity), or by product life cycle management (new models will soon replace old ones). Several partners may be involved, e.g., suppliers of service parts. If promotion planning is not sufficiently coordinated, peaks in demand may result and raw materials, components, or facilities may cause bottlenecks. Suppliers try to avoid a situation where several customers start promotions at the same time for the same product groups.	Cf. Sections 4.2.3.A and 4.3.2.C.

Desirable Features	Comments	Coverage by SAP SCM™
q) Variable set-up of parameters distinguishing between forecast-consumption and plan-consumption.	When an order arrives it is necessary to judge whether it is the realization of a planned order or not. If in the first case the order entry date differs from the planned arrival date, it has to be decided whether the incoming order is a realization of the earlier or the later planned order. If the incoming order is allocated to the later one, the system regards the order planned for a past period as an error, assumes decreasing demand, and therefore reduces the new forecast. If the order is classified as unplanned, this can be a trigger to increase forecasts and/or demand plans. This decision may be automated by a set of rules. In an SCM relationship this decision can be supported by information exchange between customers and supplier in a more systematic and efficient way than in other business relationships.	Cf. Section 4.2.1.F.
r) Definition of rules for the level of safety stocks (e.g. in distribution centers) and capacities.	The situation where partners hold too high or too low inventory levels should be avoided, because the chance to optimize the total inventory in the supply chain would be reduced if they did. *Example: If the partners know that they can access stocks of other partners if necessary they do not need to keep high safety stocks of their own. In a case study inventory positioning was identified as the by far most important driver for improving supply chain metrics (Simchi-Levi et al., 2008).*	For internal collaboration cf. Sections 4.2.2.A and 4.4.2.B. Collaboration with external partners regarding safety stock levels is not covered by SAP SCM™.

Desirable Features	Comments	Coverage by SAP SCM™
s) Backorder processing with different options.	If backorders occur, the confirmation of customer orders may be temporarily revoked for planning purposes. The option selected will influence the deployment process and affect which orders are fulfilled in-time.	Cf. Section 4.2.5.G.
t) Formulation of rules for allocating scarce products to distribution centers and warehouses in case of shortages (Deployment).	The system should support consistent behavior by providing deployment rules.	For details of deployment rules implemented in SAP SCM™ see Sections 4.2.2.I and 4.4.4.A.
u) Manifold simulations to evaluate alternative procedures.	To be considered: retrospective forecasts together with alternative set-ups of the parameters of the forecasting systems and of forecast-consumption methods [see items g) and q) above], for the safety stocks [see item r)], and for deployment [see item t)].	Cf. Section 4.2.

2.3.3.2 Operational Sales and Demand Planning

Desirable Features	Comments	Coverage by SAP SCM™
a) Methods for ATP (Available-to-Promise) and CTP (Capable-to-Promise).	With the ATP procedure the system checks whether a date and a quantity specified by a customer can be confirmed by the supplier. Only inventory and planned shipments are considered. CTP (Capable-to-Promise) assumes that the customer asks which quantity can be supplied at which date. If the customer accepts, CTP generates a new order (procurement, production, and/or transportation) to cover the new demand. Capacity and time constraints are taken into account.	Cf. Section 4.2.5.
b) Query of stocks and production orders across the borders of a firm, of capacities of the suppliers of raw material, parts and (transportation) services both in ATP and in CTP checks. Consideration of the results with respect to own operations. Include statistical values for scrap. Priority rules to choose from alternative actions when problems with customers' requests arise. Evaluation of alternative solutions (e.g., stock transfer by cargo flight versus sale of more expensive products at a reduced price).	Sometimes there are problems even within a big enterprise with decentralized stock keeping in determining where products and components are available. This may be caused by incompatible IT systems. The problem is even harder to solve in supply networks in which external partners are involved. One way out is to have common portals where the local systems can see which and how many items are available in different locations.	Only within the corporate group; cf. Section 4.2.5.B.

Desirable Features	Comments	Coverage by SAP SCM™
c) Links between final products and materials with their substitutes, together with information about potential advantages and disadvantages of a substitution.	Information about well-suited substitutes is especially difficult to obtain in internationally organized supply chains, since product features concerning quality, colors, durability, prize, etc. may be differently evaluated in different countries.	Cf. Section 4.4.6.
d) Classification of customers and orders; connecting the classes with priorities for order acceptance and execution.	Support of decision making in resource conflicts.	For backorder processing cf. Section 4.2.5.G.
e) Priorities for the use of scarce production resources such as machine tools.	Precondition: Alternative bills of materials and routings have been documented in the master data.	Cf. Sections 4.2.2.D, 4.2.2.E, 4.2.2.F, and 4.2.4.D.
f) Selection of regional, national, and global transportation facilities.	Trade-off between cost, time, and environmental impact of transportation options.	Cf. Section 4.2.6.A.
g) Consideration of pick-up windows.	Pick-up windows determine when customers or carriers are allowed to pick up products at the suppliers' site. They are an important restriction for transportation planning.	Cf. Section 4.2.6.
h) Using geo data to localize transports and to forecast arrival times.	May be based on RFID techniques.	Geo data is used to calculate transportation durations; cf. Section 4.2.6.A.
i) Manifold simulations to evaluate alternative procedures.	Simulations may help to analyze the effects of alternative priority rules; see items b) and e) above.	Cf. Section 4.2 (but not in combination with backorder processing as described in Section 4.2.5.G).

2.3.4 *Procurement Planning*

2.3.4.1 **Strategic Procurement Planning**

Desirable Features	Comments	Coverage by SAP SCM™
a) Make-or-buy decisions.	Portfolio models and Linear Programming models may be used to determine the consequences of different extents of vertical integration.	Not in scope of SAP SCM™.
b) Decisions between centralized and decentralized procurement.	Desktop purchasing systems may provide transparency even if MRO parts are ordered decentrally.	Not in scope of SAP SCM™.
c) Decisions between single sourcing and multiple sourcing of certain parts.	Effects of competition, risks of unavailability, and the advantages of close cooperation with preferred suppliers have to be considered.	Not in scope of SAP SCM™.
d) Methods to evaluate suppliers of products and services.	These methods may help in selection of partners for a long-time collaboration in supply chains and thus in structuring the network. Evaluating the technological and financial position of the suppliers may be as relevant as statistical data on past performance. Multiple-criteria decision-making procedures may be used to study the trade-offs between different procurement goals.	Not in scope of SAP SCM™.
e) Contract management.	Archiving contracts and defining alert mechanisms when contracts may be prolonged, cancelled, or modified.	Not in scope of SAP SCM™.

2.3.4.2 Tactical Procurement Planning

Desirable Features	Comments	Coverage by SAP SCM™
a) Directory of sources.	Global directories of sources for raw materials, parts, products, and services are needed when suppliers have to be replaced rapidly.	Not in scope of SAP SCM™.
b) Administration of sourcing priorities and quota arrangements.	Rules for sourcing decisions should be defined and documented.	Not in scope of SAP SCM™.
c) Coordination of safety stock policies; visualization of interdependencies between safety stocks for different components, products, and locations.	It is necessary to decide to what extent safety stocks in warehouse A can be used when there is an urgent demand in warehouse B, and vice versa. Advantages of lower safety stocks have to be compared with disadvantages of stock transfers between warehouses.	Cf. Section 4.4.4.B.
d) Simulation models.	May be helpful for analyzing safety stock policies; cf. item c) above.	Cf. Section 4.2.

2.3.4.3 Operational Procurement Planning

Desirable Features	Comments	Coverage by SAP SCM™
a) Call for tenders on B2B platforms; organization of (reverse) auctions.	For components with uncritical properties it is also important in supply chains to select the best offers.	Not in scope of SAP SCM™.
b) Pegging.	Assignment of customer orders to production orders, procurement orders, transportation orders, and other sources.	Cf. Section 4.2.4.A.

Desirable Features	Comments	Coverage by SAP SCM™
c) Collection of tracking data, considering transports on their way for planning, scheduling, and accounting.	Real-time data about the location of materials is important for short-term planning and scheduling.	No direct link from tracking data outside the boundary of the corporate group to planning.

2.3.5   Production Planning

2.3.5.1   Strategic Production planning

Desirable Features	Comments	Coverage by SAP SCM™
a) Coordination of the production capacities between the partners in the supply chain, especially between factories that can produce identical items and thus support each other in the case of bottlenecks.	In some cases the coordination is organized as an internal market on which the plants offer their products.	Not in scope of SAP SCM™.
b) Considering subcontracting and outsourcing as way out of capacity bottlenecks.	If the software does not support such decisions directly, the user might include the additional transports as set-up operations/costs.	Cf. Section 4.2.4.G for operative subcontracting.
c) Manifold simulations to evaluate alternative actions.	Simulations may support the choice of the alternatives; cf. items a) and b) above.	Cf. Section 4.2.

2.3.5.2 Tactical Production Planning

Desirable Features	Comments	Coverage by SAP SCM™
a) Cross-plant planning.	In supply chains with tight supplier-customer relations it is worthwhile to plan and schedule the capacities of plants in a similar way to those of machines and workplaces in an intra-plant MRP system. Transportation times between the plants would then correspond to set-up times in an MRP system.	Cf. Section 4.2.2.
b) Rules for access to safety stocks and for stock transfer of semi-finished products when shortages occur.	Agreements between the partners in a supply chain should include such rules to avoid complicated negotiations in critical situations.	Available within company. Not available between different companies; cf. Section 4.4.4.
c) Coordinated production and distribution planning.	Especially in global supply chains, if products have a short life cycle and distinct seasonal peaks, it is challenging to take into account the restrictions (time and quantity) for the downstream distribution. In this case the production plan is derived from the distribution plan. The theoretical optimum would be achieved by simultaneous planning, but in practice this is too complex in many cases.	Cf. Section 4.2.2.
d) Flexible definition of resources, e.g., minimal and maximal load, calendar-dependent capacities.	Typical examples for calendar-dependent capacities are reduced openings because of vacation closedowns or scheduled maintenance tasks.	Can be defined in master data; cf. Section 4.7.3.

Desirable Features	Comments	Coverage by SAP SCM™
e) Calculations of additional costs for a short-term enlargement of capacities, e.g., by running overtime.	Easing of a bottleneck in one factory may result in benefits for several partners. Therefore the contribution of each of the partners should be calculated with methods that are standardized in the supply chain to avoid recurrent negotiations.	Cf. Section 4.2.2.E.
f) Planning without final assembly.	Often the final assembly is not initiated until after an order comes in from the customer. However, "virtual intermediate products" may be produced and stored without order because they are used in several final products and will reduce the throughput-time.	Cf. Section 4.2.1.G.
g) Subassembly planning.	The assembly of the components is planned according to consumption. When customers' orders arrive, the components are "consumed." In contrast to item f), one does not plan production of the components only, but also the assembly.	Cf. Section 4.2.1.G.
h) Definition and application of different priority rules for Capable-to-Match (CTM).	Several production orders may compete for scarce resources. When a CTM analysis is applied, priority rules are necessary to determine realistic delivery dates.	Cf. Section 4.2.2.F.
i) Manifold simulations to evaluate alternative decisions.	Simulation may support the decisions mentioned in items b), c), and h) above.	Cf. Section 4.2.

2.3.5.3 Operational Production Planning and Scheduling

Desirable Features	Comments	Coverage by SAP SCM™
a) Production scheduling agreements.	These agreements are very important in industries where set-up times are long and set-up costs high. If the production plans are not harmonized and JIT delivery is agreed on, it may happen that the supplier has to produce a big lot when it does not fit into the set-up cycle because the customer urgently needs the component for its own production process.	Exchange regarding net requirements as a result of MRP; cf. Sections 4.2.3.C, 4.2.3.D, 4.3.1.B, and 4.3.1.C.
b) Heuristics for lot sizing and for priority rules, considering bottlenecks and impact on suppliers and customers.	In tactical production planning lot sizes are determined on a rough basis. These lot sizes may be modified in detailed planning and scheduling to take the actual situation in the production environment into account.	Lot sizes are regarded as master data; cf. Section 4.7.3. Heuristics for determining lot sizes are also available.
c) Alert if the synchronization of processes is violated.	Alerts when production and distribution processes which originally had been planned collaboratively are no more synchronized because of rescheduling measures, disturbances, or interruptions; the alerts have to be sent to all departments involved, e.g., by an inter-company workflow.	Cf. Section 4.6.2.

Desirable Features	Comments	Coverage by SAP SCM™
d) Considering aggregated scrap.	The scrap accrued within the supply chain must be considered in production and distribution planning.	Can be defined in master data; cf. Section 4.7.3.
e) Backorder processing, especially administration of priorities for customers and orders.	Determine most urgent backorders for earlier processing. Backorders must be considered in ATP and CTM.	Cf. Sections 4.2.4.C and 4.2.5.G.
f) Net change planning.	Because of the tight coupling of the procurement, production, and distribution plans within the supply chain, determining a new plan after each modification typically is uneconomical. Therefore methods of net change planning and scheduling, regarding only the modified data, are important so that provisional results can be obtained. However, certain (mostly temporal) events trigger the total update of all plans in a new planning run.	SAP systems apply net change planning in most cases.

2.3.6   Distribution Planning

2.3.6.1   Strategic Distribution Planning

Desirable Features	Comments	Coverage by SAP SCM™
a) Definition of rules for backorder processing.	Backorders may result from disturbances on the sell-side (e.g., because an urgent order has to be served), from production (e.g., because of interruption of a production line or unexpected scrap), or from the buy-side (e.g., delay in delivery). Distribution strategies should be defined to allow rush transports in case of urgent delivery needs to partially compensate for the delays that resulted in earlier stages.	SAP SCM™ allows to model the rules in Backorder Processing; cf. Section 4.2.5.G.
b) Make-or-buy decisions for transportation services.	Use of own truck fleet vs. contracting professional shippers.	Not in scope of SAP SCM™.
c) Degree in which value-added services of service providers are accepted.	Decisions about buying services offered as "Fourth-Party Logistics."	Not in scope of SAP SCM™.
d) Implementation of track-and-tracing functionalities.	May imply use of RFID instead of barcode.	Not in scope of SAP SCM™.

2.3.6.2 Tactical Distribution Planning

	Desirable Features	Comments	Coverage by SAP SCM™
a)	Package planning coordinated with clients and suppliers.	Package planning has to consider, e.g., the transport equipment. Client-specific packages can be necessary in addition to the standard package. Package planning may be integrated with product design and means of transport decisions, e.g., adaptation of boxes to vials and tubes in the cosmetics industry, bicycles without overhanging parts so that the storage space in the means of transportation is well utilized.	Not in scope of SAP SCM™.
b)	Administration of delivery windows.	Delivery windows at the customer's site are as important as pick-up windows at the supplier's site. Time slots may vary, e.g., during the holiday season.	Cf. Section 4.2.6.A.
c)	Supply distribution; push without demand.	Sometimes a factory has the power to push products to the point of sale even if there is no immediate demand.	Only in the internal supply network; cf. Section 4.2.2.F.
d)	Procedures for deployment; agreement of deployment rules.	Deployment becomes relevant if demand exceeds production capacity or scarce stocks have to be allocated. In this case different deployment procedures and priority rules must be applied, e.g., agreed quotas, precedence for filling safety stocks, allocation proportional to past deployment, and pro rata fulfillment of open orders. In refined versions transports and their costs will also be considered, so that (e.g.) a full shipload is accomplished.	Cf. Sections 4.2.2.I and 4.4.4.A.

Desirable Features	Comments	Coverage by SAP SCM™
e) Transport ATP and CTP.	By analogy with planning of delivery dates for production there is also a need to organize delivery dates for distribution. In developing distribution plans the availability of means of transport or the time and costs for delivery from other warehouses (e.g., for important spare parts) have to be considered.	Cf. Section 4.4.4.B.
f) Presentation of alternatives and/or substitute solutions.	Examples are cargo flights when waterways are temporarily impassable. Usage of faster means of transport in case of delays or when delivery of spare parts is urgent.	Cf. Sections 4.2.2.E and 4.4.4.A.
g) Determination of transport lot sizes.	Transport lot sizes can strongly influence the production and delivery planning of partners when goods have high carriage costs.	Not in scope of SAP SCM™.
h) Relocation of stocks.	In some cases it may be necessary to modify earlier distribution decisions and to change the location of certain stocks to fill urgent orders from customers or to provide sufficient safety stocks.	Cf. Section 4.4.4.B.
i) Coordination of intermediaries.	Logistics service providers are often employed for some or all distribution tasks. These companies often belong to the class of Small and Medium Enterprises (SME), for which it may be difficult to implement sophisticated IT systems for seamless collaboration with their supply chain partners.	Cf. Section 4.6.1.
j) Manifold simulations for evaluation of alternative procedures.	Simulations can be helpful, e.g., for items f) and h) above.	Cf. Section 4.2.

2.3.6.3 Operational Distribution Planning and Scheduling

Desirable Features	Comments	Coverage by SAP SCM™
a) Shelf-life monitoring.	The supplier may be responsible for shelf-life management, including the placement of articles in the shelves and observing dates of expiry, notably for food, cosmetics, and pharmaceuticals. Destruction of expired products can result in urgent distribution activities. In VMI relationships the supplier may have to consider POS data for forecasting distribution needs.	Only within company; cf. Section 4.2.4.E.
b) Consideration of tracking dates and of stocks on their way in accounting and planning.	At accounting dates (e.g., for quarterly reporting) stocks on their way have to be documented and valued. Rules are needed on whether stocks are to be committed as assets of the supplier or of the customer when they are en route.	Stocks in transit are considered in the planning modules. Accounting is not in the scope of SAP SCM™.
c) Consideration of storage and handling restrictions.	Examples are special procedures for transporting dangerous goods or special cranes in harbors.	With limitations, cf. Sections 4.2.2 and 4.2.4.F.
d) Transport leitstand.	Similar to control units ("leitstands") employed in production scheduling, transport leitstands may visualize the use of transportation resources and the dependencies between consecutive transport activities.	Cf. Sections 4.2.2.C and 4.2.6.A.
e) Use of previously agreed deployment rules.	Rules defined at the tactical level are applied in the case of shortages.	Cf. Sections 4.2.2.I and 4.4.4.A.

Desirable Features	Comments	Coverage by SAP SCM™
f) Vehicle scheduling regarding incompatibilities.	For example, security regulations dictate that the transport of hydrogen and oxygen tanks on the same vehicle must be avoided.	Cf. Section 4.2.6.A.
g) Transport load builder.	Several strategies should be represented in the software, e.g., loading of trucks with the same goods independently of the target location or loading of trucks that drive to one target location with goods of different types and with different delivery dates.	Cf. Section 4.2.2.J.

2.3.7 Return and After-Sales Processes

Desirable Features	Comments	Coverage by SAP SCM™
a) Especially in make-to-order production it is very important for the service-person to know which components have been used in producing a product which is out of order.	Detailed documentation of product properties in PDM systems is important. The service-person should have easy access to this information. Often the after-sales service is provided by a different company; in this case sharing the data is an important issue for efficient after-sales services.	Not in scope of SAP SCM™.
b) Customer care should be supported beyond the purchasing act.	For products with a typical usage time it is important to contact the customer in time to influence his replacement purchase. A close coordination with CRM systems is necessary.	Not in scope of SAP SCM™.

Desirable Features	Comments	Coverage by SAP SCM™
c) Recycling can be supported by design processes and detailed documentation of product properties in PDM systems.	The limited availability of natural resources makes recycling of components at the end of the life cycle of the associated product a high priority. This information should be shared with companies that are active in the recycling process.	Not in scope of SAP SCM™.
d) Returning handling units.	Special containers, bins, pallets, etc. may have to be returned to the supplier and a "Reverse Supply Chain" has to be organized using similar planning and scheduling methods as in the "Forward Supply Chain."	Not in scope of SAP SCM™.

2.3.8   Controlling and Support Processes

2.3.8.1   Supply Chain Event Management (SCEM)

Desirable Features	Comments	Coverage by SAP SCM™
a) Detection of events.	The more details IT systems are recognizing, the higher the potential for defining and detecting events that may be relevant internally and for supply chain partners.	Cf. Section 4.6.1.
b) Rule-based filtering of detected events using a flexible, parameter-controlled definition of what should be considered a relevant exception.	Filtering mechanisms are highly relevant to avoid information overloads.	Cf. Section 4.6.1.

Desirable Features	Comments	Coverage by SAP SCM™
c) Assignments of events/deviations to institutions and bearers of roles, who have to be informed. Information of these addressees inside and outside the firm (workflow management).	Events that are relevant internally may not be of interest for other partners in the supply chain. Different filtering parameters can be applied for different partners in the supply chain.	Cf. Section 4.6.1.
d) Simulation of the impact of events on the buy-side (upstream simulation) and on the sell-side (downstream simulation) (diagnosis).	Some situations that should be simulated are: upstream, the loading/overloading of warehouses if the suppliers cannot stop their production processes while there are distribution problems; downstream, the impairment of customers' readiness for delivery if safety stocks are violated.	Not in scope of SAP SCM™.
e) Selection of standardized remedial actions using priority rules (proposal of therapy).	Not all possible actions can be proposed by the IT system, but only those that are predictable or standardized.	Not in scope of SAP SCM™.
f) Simulation of impact of remedial actions on the supply chain as a whole (forecast of effects of therapy).	*Example: Parts reserved for customers B and C could be sent to the prioritized customer A because of an accident during transport to A. The simulation tries to determine when the production at plants B and C may have to stop because of missing materials. The impact on B's and C's customers may also be considered. What would be the effect if A were served later and the orders from B and C were fulfilled in time?*	Not in scope of SAP SCM™.

2.3.8.2 Management Information/Performance Management

Desirable Features	Comments	Coverage by SAP SCM™
a) Visualization of actual states by activating nodes and arcs.	This feature allows the selection of detailed data without showing the same level of detail in other, currently less relevant areas. *Examples are capacity usage, frequency of disturbances, or delays.*	Cf. Section 4.6.2.
b) Tracing procurement processes.	Upstream tracing over several partners in the supply chain is especially important in industries where certain problems have to be detected immediately (e.g., contamination in the food industry or mechanical ruptures in the construction industry).	SAP Event Manager™ (cf. Section 4.6.1) might cover this to some extent.
c) Methods for evaluating customers.	Methods to calculate the customer lifecycle values are needed to define priorities within deployment.	Not in scope of SAP SCM™.
d) Generator for performance measurement systems (e.g., DuPont scheme, Value driver trees, Balanced Scorecards).	Flexible generation of reports on KPIs should be provided.	Not in scope of SAP SCM™.
e) Adaptation of the metrics provided in the SCOR® model to a KPI system that is suitable for the supply chain.	For the SCOR® model and the metrics proposed in it cf. Section 2.2.2.	Not in scope of SAP SCM™, but of SAP BI™.
f) Delivery of business content, e.g., average values or benchmark data for certain industries.	The use of such data, provided by associations like the SCC, consultancies, or the producer of SCM software can contribute to continuous improvement of the supply chain.	Not in scope of SAP SCM™, but of SAP BI™.

Desirable Features	Comments	Coverage by SAP SCM™
g) Exchange of indicators with the supply chain partners in a standardized format.	*Examples: By using XML or the Extensible Business Reporting Language XBRL.*	Not in scope of SAP SCM™.
h) Balanced Scorecard for SCM. Coordination of the SCM Scorecard with other Scorecards within the firm and with those of supply chain partners.	If the goals and the indicators in the Balanced Scorecard of one firm are coordinated with those in the Balanced Scorecards of the partners a kind of competition within the supply chain may result. This is important when suppliers of similar products, components, or raw materials must be compared. The Scorecards of different partners in the supply chain may look quite different owing to industry specifics. An SCM system should provide support in aggregating and coordinating the different Scorecards. The values can be compared over time and, if relevant benchmarking data for similar supply chains is available, also with this data.	Not in scope of SAP SCM™.

Chapter 3

Processes of SCM Covered by mySAP SCM 5.0

3.1   Introduction

The purpose of this section is to explain the mySAP SCM solution by focusing on generic processes. While **mySAP SCM** describes the SCM processes that are in the scope of the SAP solution (or partner products), **SAP SCM™** describes the product. The following pages provide a survey of the mySAP Business Suite and an overview of the mySAP SCM solution. The chapter lists all generic processes enabled by mySAP SCM at a glance.

Each process is described in several aspects:

- First the business background and main aspects are described.
- Secondly the products and components that enable the process are also listed.

Generic SCM processes are building blocks for scenarios together with industry-specific processes and can be used in scenarios from various industries.

3.2   mySAP Business Suite

To complement the entire range of SCM processes within mySAP SCM, the solution uses other solutions from the mySAP Business Suite such as SAP Customer

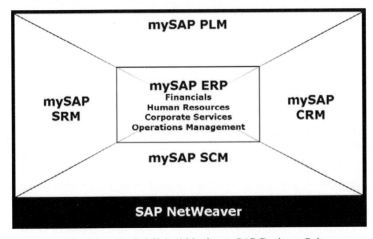

Fig. 3.1 mySAP SCM within the mySAP Business Suite

Relationship Management (SAP CRM™, e.g., for Trade Promotion Management), and SAP Supplier Relationship Management (SAP SRM™), powered by SAP NetWeaver™.

The solution map describes the processes that are supported by SAP or partner products. mySAP SCM subsumes the processes of the solution map for SCM. However, mySAP SCM is not a product (in contrast to SAP SCM™), and more than one product is required to cover the processes of mySAP SCM. For example, in addition to SAP SCM™ also SAP ERP™ and SAP SRM™ are required (cf. Fig. 3.1).

3.3   mySAP SCM

The mySAP Supply Chain Management (mySAP SCM) solution consists mainly of the SAP Supply Chain Management (SAP SCM™) and parts of SAP ERP™ (cf. Fig. 3.2).

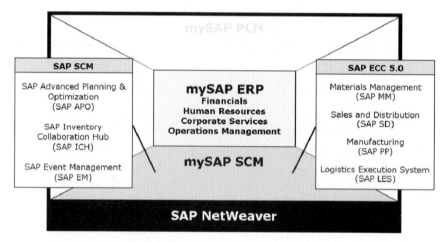

Fig. 3.2 The components of mySAP SCM

mySAP SCM consists of following subsystems:

- SAP Advanced Planning and Optimization (APO) (SAP APO™),
- SAP Forecasting and Replenishment (SAP F&R™),
- SAP Inventory Collaboration Hub (SAP ICH™), recently renamed as SAP Supply Network Collaboration (SAP SNC™),
- SAP Event Management (SAP EM™), and
- SAP Extended Warehouse Management (SAP EWM™).

Service Parts Planning is another application which uses mostly disjunctive functionality of SAP APO™ and SAP ICH™. The scope of each of these application systems is described in Section 4.1.

The mySAP SCM solution uses SAP Business Intelligence (SAP BI™) for information integration, SAP Enterprise Portal (SAP EP™) for people integration, and SAP Exchange Infrastructure (SAP XI™) for process integration.

3.4 Generic Processes Enabled by mySAP SCM 5.0 at a Glance

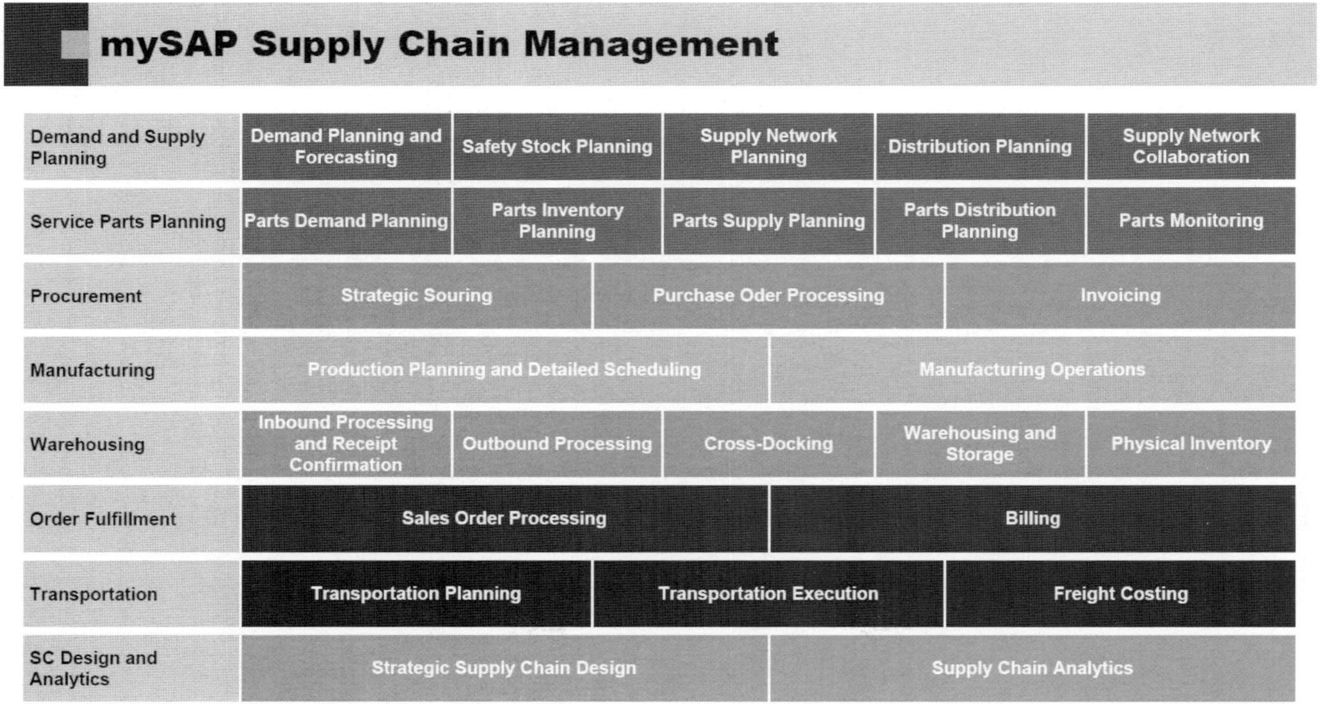

mySAP Supply Chain Management

Demand and Supply Planning	Demand Planning and Forecasting	Safety Stock Planning	Supply Network Planning	Distribution Planning	Supply Network Collaboration
Service Parts Planning	Parts Demand Planning	Parts Inventory Planning	Parts Supply Planning	Parts Distribution Planning	Parts Monitoring
Procurement	Strategic Souring		Purchase Oder Processing		Invoicing
Manufacturing	Production Planning and Detailed Scheduling			Manufacturing Operations	
Warehousing	Inbound Processing and Receipt Confirmation	Outbound Processing	Cross-Docking	Warehousing and Storage	Physical Inventory
Order Fulfillment	Sales Order Processing			Billing	
Transportation	Transportation Planning		Transportation Execution		Freight Costing
SC Design and Analytics	Strategic Supply Chain Design			Supply Chain Analytics	

Fig. 3.3 Generic processes based on mySAP SCM 5.0

3.5   Process View of mySAP SCM

3.5.1   Demand and Supply Planning

Process	Comment
1. Demand planning and forecasting	Demand planning predicts the anticipated customer demand at a finished product level. ■ Integration of any data ■ Analysis of historical data ■ Forecasting ■ Life cycle planning ■ Promotion planning ■ Flexible planning board and graphics ■ User specific settings and data assignment ■ Consistent planning on any level ■ Any kind of calculations by macros ■ Offline planning ■ Analytics – for forecast accuracy, for example alerts. Demand Planning and Forecasting requires the following products and components: ■ Products: SAP SCM™ ■ Components: Demand Planning (APO-DP; see Section 4.2.1).
2. Safety stock planning	Establish the appropriate level of safety stock inventory for all intermediate and finished products at their respective locations to meet a target service level. ■ Quantity-based and period-based definition of service level ■ Dynamic and multistage calculations of safety stock ■ Considers multilevel distribution networks ■ Considers multilevel bills of material ■ Considers demand variability, supply variability, lot sizes, and safety horizon ■ Inventory and service level analytics (SCOR$^{®}$). Safety Stock Planning requires the following products and components: ■ Products: SAP SCM™ ■ Components: Supply Network Planning (APO-SNP; see Section 4.2.2).
3. Supply network planning	■ Supply network planning propagates demand information through the supply network and determines the sources of supply to fulfill independent and dependent requirements. ■ Integrates purchasing, production, distribution, and transportation for • reduced order fulfillment times • reduced inventory levels • improved customer service ■ Simulation and implementation as comprehensive tactical planing and sourcing decisions ■ Considers constraints and penalties to plan product flow along the supply chain ■ Enables planning on different levels of details (aggregated planning) ■ Enables service, inventory, and asset utilization analytics. Products: SAP SCM™, SAP BI™ ■ Components: Supply Network Planning (APO-SNP; see Section 4.2.2).

Process	Comment
4. Distribution planning	Distribution Planning determines the best short-term strategy to allocate available supply to meet demand and to replenish stocking locations. • Determines which demand can be fulfilled by the existing supply • Depending upon the result, the supply network plan will be confirmed or changed and a plan for stock transfers will be created • Rule-based transport load building. Distribution Planning requires the following products and components: • Products: SAP SCM™ • Components: Supply Network Planning (APO-SNP; see Section 4.2.2).
5. Supply network collaboration	Support multi-enterprise business processes between suppliers and customers. • Innovative collaboration platform. • Offers VMI functionality for the replenishment of Retailer warehouses taking promotions and direct deliveries into consideration. • Supplier collaboration includes ways for customers to share, synchronize, and collaborate with suppliers on demand, inventory, and order data. Supply Network Collaboration requires the following products and components: • Products: SAP SCM™, SAP ERP™, SAP BI™ • Components: Supply Chain Collaboration (APO-COL; see Section 4.2.3), Supplier Collaboration (ICH-SUP; see Section 4.3.1), Responsive Replenishment (ICH-RR; see Section 4.3.2).

3.5.2 Service Parts Planning

Process	Comment
1. Parts demand planning	Create a demand forecast based upon historical and comparative data • Scaling of the demand to equal number of workdays and aggregation along the structure of the supply network • Evaluation of existing forecast model based upon forecast error • Automatic selection of the best-fit forecast model and its forecast parameters • Forecast models for slow-moving and sporadic demand • Demand aggregation and forecast disaggregation capabilities (along the structure of the supply network) • New parts forecasting and long-term forecast • Exception-driven forecast approval • Simple and complex supersession of parts. Parts Demand Planning requires the following products and components: • Products: SAP SCM™ • Components: Forecasting (SPP-FCS; see Section 4.4.1).

Process	Comment
2. Parts inventory planning	Determine the most efficient stocking locations and optimal inventory policies ■ Stocking and destocking decisions determine where to stock a given part (authorized stocking list determination) ■ Calculate safety stock and suggested order quantity in the network toward suppliers ■ Use safety stock calculation rules for normal distributed or Poisson-distributed demand ■ Safety stock is constant or time-phased ■ Consider virtual consolidation locations for areas with small demand ■ Support both push-driven and pull-driven material deployment ■ Identify surplus inventory and determine efficient scrap and obsolescence proposals ■ Determine expected demand and safety buffer for the remaining expected lifetime, considering minimum legal retention periods ■ Determine global or regional surplus inventory ■ Distribute surplus proposals among inventory-holding locations to handle scrap efficiently ■ Efficient surplus approval workflow ■ Provide obsolescence proposals based upon remaining parts demand ■ Analyze open and approved surplus value. Parts Inventory Planning requires the following products and components: Products: SAP SCM™ Components: Inventory Planning (SPP-INVP; see Section 4.4.2), Surplus & Obsolescence Planning (SPP-SOB; see Section 4.4.5).
3. Parts supply planning	Determine efficient supply plans and release them to the supplier ■ Determine requirements throughout the network and aggregate toward supplier-facing locations, considering optimal order quantities and rounding rules ■ Create schedule or purchase requisitions for one or multiple suppliers, considering schedule stability rules ■ Level supply plan based upon seasonal safety stock shift, supplier shut-downs, and anticipated demand coverage ■ Use-up of the predecessor product's inventory by demand for the successor product in a supersession relationship ■ Authorize and release supply plans. Parts Supply Planning requires the following products and components: ■ Products: SAP SCM™ ■ Components: Distribution Requirements Planning (SPP-DRP; see Section 4.4.3).
4. Parts distribution planning	Parts Distribution Planning distributes available parts within the network via push or pull deployment and inventory balancing. ■ Determination of demand that can be fulfilled by the existing supply ■ Creation of feasible stock transports triggered by goods receipt ■ Fair share calculation and demand prioritization ■ Lateral stock transfers based on cost-benefit analysis. Parts Supply Planning requires the following products and components: ■ Products: SAP SCM™ ■ Components: Deployment and Inventory Balancing (SPP-DEPL, see Section 4.4.4).

Process	Comment
5. Parts monitoring	Monitors the current state of the service supply chain. • Quickly identify shortages and potential shortages to avoid implications for service • Provide global inventory visibility through service parts planning cockpit • Drilldown capabilities to detailed part information • Gather all alerts related to service parts planning at a central location • Enable collaborative resolution of alerts with supplier. Parts Monitoring requires the following products and components: • Products: SAP SCM™ • Components: Alert Monitoring.

3.5.3   Procurement

Process	Comment
1. Strategic sourcing	Identification of core suppliers which are chosen for strategic relationships and definition of the parameters that drive procurement execution. • Vendor analysis and purchasing statistics to evaluate potential suppliers • Contracts and scheduling agreements • Quota arrangements for rule definition of demand distribution across sources of supply • Source lists and priorities • Performance management through cost controlling and contract compliance analytics. Strategic Sourcing requires the following products and components: • Products: SAP ERP™, SAP SRM™, SAP BI™ • Components: Material Management (ERP-MM).
2. Purchase order processing	Fulfills direct procurement requirements through the sourcing, issuance, and confirmation of purchase orders. • Purchase requisitions created by supply planning or various departments define the workload of the purchasing department • Worklist-based functions support efficient conversion of purchase requisitions to purchase orders or requests for quotation • Process variants for subcontracting, consignment, or service procurement • Special tools and functions allow detailed order status tracking and monitoring • Performance management through spend and contract compliance analytics. Purchase Order Processing requires the following products and components: • Products: SAP ERP™, SAP BI™, SAP SRM™, SAP ICH™ • Components: Material Management (ERP-MM).

Process	Comment
3. Invoicing	Receives and enters a vendor's invoice, and checks for correctness. ■ Invoice is blocked for payment if the tolerances (e.g., concerning the quality or the shipment dates) defined by the company have been exceeded. For blocked invoices, a dispute process can be initiated with the vendor. ■ After the verification step, the necessary postings are triggered in accounting. ■ Instead of entering invoices sent by the vendor, you can create the invoice based upon the purchase order prices and the received goods. ■ Invoices are the basis for payment. Invoicing requires the following products and components: ■ Products: SAP ERP™ ■ Components: Material Management (ERP-MM).

3.5.4 Manufacturing

Process	Comment
1. Production planning and detailed scheduling	Supports the process of assigning production orders to resources in a specific sequence and time ■ Matches supply to demand. Determines how, when, and where resources should be deployed to meet production goals. ■ Modeling of production conditions like setups. ■ Interactive or automated scheduling. • Interactive simulation capabilities. • Automated scheduling creates an improved, executable, capacity- and constraint-based production plan. ■ Advanced scheduling tools and solvers. Production Planning and Detailed Scheduling requires the following products and components: ■ Products: SAP SCM™ ■ Components: Production Planning and Detailed Scheduling (APO-PP/DS; see Section 4.2.4).
2. Manufacturing execution	Manufacturing execution enables manufacturers to meet and deliver on their production plans by managing production processes, workforce deployment, and resource deployment on the shop floor and to document, monitor, and dispatch inventory across the production life cycle. It supports discrete, process, repetitive, and continuous manufacturing processes. ■ Allocate and match labor, materials, and resources to orders. ■ Track customer orders, inventory, and WIP across the production life cycle. Manufacturing Operations requires the following products and components: ■ Products: SAP ERP™ ■ Components: Production Planning (ERP-PP).

3.5.5 Warehousing

Process	Comment
1. Inbound processing and receipt confirmation	Comprises all steps of an external procurement process that occur when the goods are received. ■ Advanced Shipping Notifications (ASN) ■ Registration of trucks (upon arrival at the warehouse) ■ Yard management and appointment scheduling ■ Goods receipt posting including receipt from production ■ Quality check ■ Performance management through vendor performance analytics. Inbound Processing and Receipt Confirmation requires the following products and components: ■ Products: SAP ERP™, SAP BI™, SAP EWM™ ■ Components: ERP-MM, Logistics Execution System (ERP-LES).
2. Outbound processing	Prepares and ships goods to their destination. ■ Creating outbound deliveries ■ Creating warehouse orders for picking the goods ■ Picking, packing, and value-added services ■ Staging and loading ■ Goods issue posting ■ Outbound vehicle management ■ Performance management through delivery analytics (SCOR®, for example). Outbound Processing requires the following products and components: ■ Products: SAP ERP™, SAP EWM™, SAP BI™ ■ Components: ERP-LES.
3. Cross-docking	Processes merchandise in a distribution center or warehouse where the goods are brought from the goods reception directly to goods issue without being stored. ■ Planned cross-docking: match Inbound to Outbound deliveries ■ Opportunistic (unplanned) cross-docking: detecting opportunities for cross-docking in a dynamic environment. Cross Docking requires the following products and components: ■ Products: SAP ERP™, SAP EWM™ Components: ERP-MM, ERP-LES.
4. Warehousing and storage	Processes warehouse internal movements and storage of materials. ■ Matching warehouse task and resources over different days (optimization in a warehouse) ■ Controlling warehouse movements of materials on a storage bin level ■ Depicting the physical structure of a warehouse ■ Storage and stock management for bin stocks. Warehouse and Storage requires the following products and components: ■ Products: SAP ERP™, SAP EWM™ ■ Components: ERP-LES.

Process		Comment
5.	Physical inventory	Supports all activities for planning and executing physical inventory.

- The physical inventory process can be planned (which material has to be counted at which storage location on what date)
- Stocks can be blocked for goods movements before the beginning of counting, measuring, and weighting
- When results have been entered, differences between the quantities and values can be reported.

Physical Inventory requires the following products and components:
- Products: SAP ERP™, SAP EWM™
- Components: ERP-MM, ERP-LES.

3.5.6   Order Fulfillment

Process		Comment
1.	Sales order processing	Allows order entry, pricing, and scheduling for fulfillment.

- Determination of sources, price, and partners
- Order scheduling
- ATP check (order promising)
- Routing guide (transportation feasibility)
- Pricing
- Foreign trade Global Trade System (GTS)
- Performance management through SCOR® delivery and service level analytics and sales productivity analytics.

Sales Order Processing requires the following products and components:
- Products: SAP SCM™, SAP ERP™, SAP BI™
- Components: Global Available-to-Promise (APO-ATP; see Section 4.2.5), Sales and Distribution (ERP-SD).

| 2. | Billing sales order processing | Considers all activities from issue of invoice to incoming payment. |

- Integration with financial accounting and controlling
- Invoices cr eated based on deliveries or services
- Comprehensive pricing functions
- Pro forma invoices
- Credit and debit memos
- Incoming payment.

Billing Sales Order Processing requires the following products and components:
- Products: SAP ERP™
- Components: ERP-SD.

3.5.7 Transportation

Process		Comment
1.	Transportation planning	Creates an improved and executable transportation plan.
		■ Collaborative capacity planning with carriers
		■ Automated planning is based upon respective service requirements and capacity constraints with minimum investment
		■ Supports load consolidation, order splitting based upon resources such as vehicle scheduling
		■ Selection of transportation service provider
		■ Tendering and tender monitoring
		■ Delivery proposal for synchronized picking processes delivery window
		■ Performance management through supply chain logistics costs and delivery analytics.
		Transportation Planning requires the following products and components:
		■ Products: SAP SCM™
		■ Components: Transportation Planning and Vehicle Scheduling (APO-TP/VS; see Section 4.2.6).
2.	Transportation execution	Complete and integrated solution to create, execute, and monitor shipments.
		■ Create shipments
		■ Create shipment legs
		■ Schedule
		■ Assign carriers
		■ Print documents for shipping.
		Transportation Execution requires the following products and components:
		■ Products: SAP ERP™
		■ Components: ERP-LES.
3.	Freight costing	Calculates and settles freight costs.
		■ Freight cost calculation based upon actual shipments and freight rates
		■ Basis for
		● verifying invoices sent from the carrier
		● self-billing transportation service providers (TSP)
		■ Freight cost settlement
		■ Provides visibility across the transportation execution process for international inbound and outbound shipments.
		■ Cover milestones and status information for road and sea transportation including customs issues
		■ Monitor and control complex transportation networks
		■ Send event messages for expected and unexpected events.
		■ Automatically trigger follow-up processes such as rescheduling in the case of a delay
		■ Role-based access for customer (sales executive) and forwarding agent
		■ Provides KPIs on adherence to planned durations and customs processing.
		Freight Costing requires the following products and components:
		■ Products: SAP ERP™
		■ Components: ERP-LES.

3.5.8   Supply Chain Design and Analytics

Process		Comment
1.	Strategic supply chain design	Design, evaluate, and optimize the supply chain model. Strategic Supply Chain Design requires the following products: SAP SCM™.
		For the optimization of a network, SAP recommends software from a certified partner, LogicTools (www.logic-tools.com).
2.	Supply chain analytics	Create insights across all functions. More than 1,000 pre-defined KPIs (many of them based upon the SCOR® model; cf. Section 2.2.2) for
		<ul><li>Planning</li><li>Procurement</li><li>Manufacturing</li><li>Order fulfillment</li><li>Inventory management</li><li>Visibility.</li></ul>
		Supply Chain Analytics requires the following products and components:
		<ul><li>Products: SAP SCM™, SAP BI™, SAP ERP™</li><li>Components: Event Management (SAP EM™).</li></ul>

3.6   Relationship between the Supply Chain Pyramid and the Solution Map

In Section 2.2.3 we proposed a Supply Chain Pyramid to visualize different tasks that are relevant for Supply Chain Management. Section 3.4 introduced the mySAP SCM Solution Map and in the previous Section 3.5 we described the generic processes defined in it in more detail. Based on this information we are able to visualize in Fig. 3.4 which parts of the framework are supported by SAP SCM systems and components.

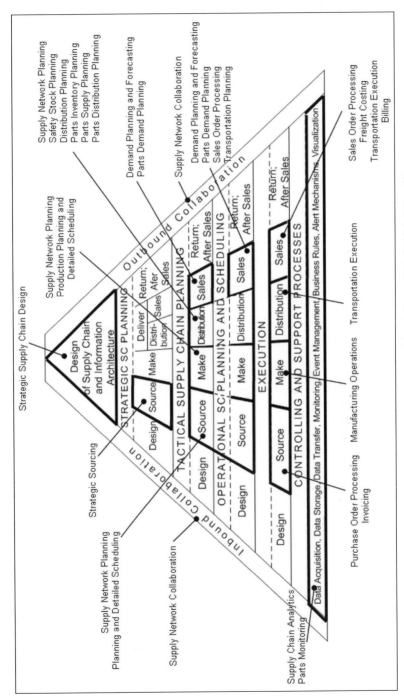

Fig. 3.4 Projection between the SCM pyramid and the mySAP SCM Solution Map

Chapter 4

SAP Systems for Supply Chain Management

4.1 Overview of SAP SCM™

The scope of this chapter is the description of the processes and functions of SAP SCM™ – which cover tactical and operational supply chain planning and collaboration – because these are the most typical processes for SCM. Another reason for the restriction to SAP SCM™ is that especially the description of the execution processes in SAP ERP™ would far exceed the scope of this book. The second restriction is that warehousing also does not fall within the scope of this book. Since warehousing is also a part of execution, we have decided to skip this topic in favor of allocating more room for supply chain planning and collaboration.

(A) Structure of SAP SCM™
In order to increase the transparency about the structure of SAP SCM™ we will use the terms product, application system and module or component for the different levels (in the style of SAP). The product SAP SCM™ is the software code that is available on DVD and installed on an instance. The product SAP SCM™ contains the following application systems:

- SAP Advanced Planning and Optimization (SAP APO™) for planning the internal supply chain and collaboration with external supply chain partners,
- SAP Inventory Collaboration Hub (SAP ICH™) recently renamed as SAP Supply Network Collaboration (SAP SNC™), for collaboration with suppliers and customers
- SAP Forecasting and Replenishment (SAP F&R™) for replenishment planning in the high volume retail industry
- SAP Event Management (SAP EM™) for tracking and tracing orders and order chains, and
- SAP Extended Warehouse Management (SAP EWM™) for the (decentralized) management of warehouses.

If SAP SCM™ is installed, all these application systems are available in the same instance as shown in Fig. 4.1. Nevertheless, separate instances are often used, e.g., for safety reasons (especially for collaboration using SAP ICH™) or for administrative reasons (especially for the management of decentralized warehouses with SAP EWM™).

There is no default integration between the application systems provided by SAP, and if integration is required, this has to be done by system integrators. Some of the application systems contain modules that group functionality for specific tasks – an example is the module Demand Planning within SAP APO™. Other modules within SAP APO™ are Supply Network Planning, Production Planning and Detailed Scheduling, Global Available-to-Promise, and Transportation Planning and Vehicle Scheduling.

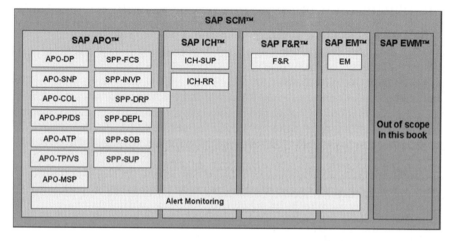

Fig. 4.1 Components of the SAP SCM™ product

In most cases the modules of an application system are integrated with each other by default. We introduce the term component because it makes sense to group a set of functions even when the functions do not belong to a module – for example the collaboration functionality in SAP APO™. Therefore a component corresponds either to a module or contains a set of functions working together pursuing a certain purpose.

(B) Advanced Planning and Optimization

The Advanced Planner and Optimizer SAP APO™ was released in 1998 as the first SCM application system, and it still offers the most extensive functionality. The planning components of the application system SAP APO™ are:

- Demand Planning (APO-DP) for planning and forecasting customer demand,
- Supply Network Planning (APO-SNP) for distribution planning and replenishment of the internal supply chain and for as rough-cut production planning,
- Supply Chain Collaboration (APO-COL) for collaborative processes with external supply chain partners, e.g., Vendor Managed Inventory,
- Production Planning and Detailed Scheduling (APO-PP/DS) for advanced planning and scheduling at plant level,
- Global Available-to-Promise (APO-ATP) for confirmation and sourcing of (mostly customer) demands,

- Transportation Planning and Vehicle Scheduling (APO-TP/VS) for scheduling and tendering of transports and combination of freight units, and
- Maintenance and Service Planning (APO-MSP), with industry focus on the airline industry.

(C) Inventory Collaboration Hub

The Inventory Collaboration Hub SAP ICH™ was first released in 2003 as the second application system for SCM. In recent releases its name changed to Supply Network Collaboration (SAP SNC™), and this application system will also be available separately (i.e., without the context of SAP SCM™). The functionality for supplier collaboration (ICH-SUP) enables industry-independent collaboration with supplier via the Internet, while the functionality for customer collaboration – Responsive Replenishment (ICH-RR) – is designed to support the consumer product industry.

(D) Forecasting and Replenishment

The purpose of SAP F&R™ is a retail industry specific application system with the purpose of planning the short-term replenishment of the very high number of SKUs in a retailer's stores and distribution centers. SAP F&R™ uses some master data of SAP APO™ but is designed as an alternative to replenishment planning in SAP APO™.

(E) Event Management

SAP EM™ is used for monitoring supply chain processes within and outside the boundaries of the company. Therefore its connectivity to other application systems is a key feature of SAP EM™, and consequently any combination with other application systems is possible.

(F) Service Parts Planning

For the specific requirements of service parts management (e.g., sporadic demand) a solution has been developed on the basis of SAP ERP™, SAP CRM™, and SAP SCM™ (Dickersbach 2007). The entire service parts management (SPM) solution contains such scenarios as "procure to stock," "sell from stock," "third-party order processing," "claims and returns," and "entitlement management." The SPM solution is structured into service parts planning (SPP) and service parts execution (SPE). The planning part of this solution contains the components:

- Forecasting (SPP-FCS) for the capturing and management (interactive adjustment) of historical demand and forecasting of future demand,
- Inventory Planning (SPP-INVP) for the determination of inventory levels and order quantities,
- Distribution Requirements Planning and Procurement (SPP-DRP) to determine order quantities for external procurement,
- Deployment and Inventory Balancing (SPP-DEPL) to replenish the internal supply network,

- Surplus and Obsolescence Planning (SPP-SOB) to determine and remove surplus quantities from the supply network, and
- Supersession (SPP-SUP) for the replacement planning of one part by another.

Though the functionality for SPP is technically located in SAP APO™ for planning and in SAP ICH™ for monitoring, as shown in Fig. 4.1, the functionality is almost disjunctive, and mixing with the functionality with the "standard" SAP APO™ is not intended. Therefore, we will refer to SPP as a separate application system.

(G) Coverage of the Solution Map by SAP SCM™ Components

Fig. 4.1 provides an overview of the SAP SCM™ components that are in the scope of this chapter by application system. Service Parts Planning is based technically mainly on SAP APO™, but it also requires SAP ICH™ for purchase order collaboration and alert monitoring.

Introducing the SAP SCM™ components also helps in determination of the coverage of the mySAP SCM Solution Map by the product SAP SCM™. Fig. 4.2 shows how the mySAP SCM processes are covered by the SAP SCM™ SCM components.

The Solution Map contains strategic procurement and procurement execution. The rather poor coverage of the Solution Map by SAP SCM™ in this area is because most of its procurement functionality is tactical and operational procurement planning (in the components APO-SNP, APO-PP/DS, and SPP-DRP). However, the Solution Map does not contain tactical and operational procurement planning.

(H) Generic Functionality and Industry-Specific Requirements

Most of the functionality in SAP SCM™ is generic in the sense that it is not specific for any one industry. Most of the SAP SCM™ customers belong to the consumer products, high tech, mill products, chemicals and life science, industrial machinery and components, automotive, wholesale, retail, aerospace, oil and gas, and media industries – there are even SAP SCM™ customers for banking. Though these industries have very different processes and requirements, the flexibility of SAP SCM™ allows to model at least the more common ones. Nevertheless, there are some components that focus on a specific industry:

- Forecasting and Replenishment in SAP F&R™ for retail industry,
- Responsive Replenishment in SAP ICH™ for the consumer products industry, or
- Maintenance and Service Planning in SAP APO™ for the aerospace industry.

In the APO PP/DS component, the dependence on the physical production processes requires industry-specific functionality. These requirements are addressed (e.g.) by repetitive manufacturing for automotive, shelf life and campaign planning for process industry, project manufacturing and variant configuration for industrial machinery and components, and the configure-to-order scenario for mill products.

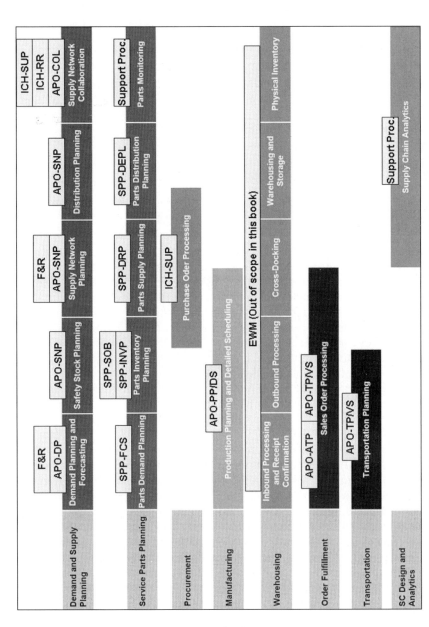

Fig. 4.2 Coverage of the mySAP SCM Solution Map by SAP SCM™ components

(I) Functionality and Application Systems

Similar functionality exists more than once within the application systems of SAP SCM™ – e.g., forecasting is supported by APO-DP, ICH-RR, SPP-FCS, and SAP F&R™. Especially in the area of service parts planning there is a huge overlap with Demand Planning and Supply Network Planning in SAP APO™. The reasons for this lie in industry-specific or after-sales-specific requirements.

This chapter is structured by application systems. The idea of structuring it by processes was discarded because in most cases the functionality of an SCM component has to be regarded in the context of its application system. For example, it is not possible to combine forecasting from SAP F&R™ and distribution planning from APO-SNP (at least not without additional development effort). Another advantage of the application system-related structure is that the intended use of the SAP SCM™ components becomes more clearly visible.

4.2   Supply Chain Management with SAP APO™

SAP APO™ focuses on the SCM within a company (i.e., collaboration with internal supply chain partners) but also provides some collaboration functionality with external supply chain partners. SAP APO™ uses a persistent memory for planning and separate engines for optimization tasks, and contains the modules listed in Section 4.1.B. The functions for collaboration are partially assigned to these modules, e.g., collaborative demand planning to Demand Planning or VMI to SNP. In order to provide a concise and central overview of the collaboration functionality, for the purposes of this book we have grouped these functions under the separate component Supply Chain Collaboration (APO-COL).

The processes span different time horizons, from long-term demand planning to short-term transportation planning. Therefore, the planning granularity is also different: in DP and SNP time buckets of months, weeks or days are used, in ATP usually of days, while PP/DS and TP/VS apply a time-continuous calculation, so that all orders are scheduled to hour, minute, and second.

Technically, SAP APO™ contains three parts for data storage: the database, the SAP BI™ data mart, and the liveCache. The liveCache is basically a huge main memory where planning and scheduling is performed and the relevant data is kept to enhance the performance for complex calculations.

The two basic structural elements in SAP APO™ are the model and the version. Several versions can be assigned to one model. The general idea is that the model contains the master data and the version the transactional data, but for some master data (location, product, and resource) it is also the possible to make some version-dependent changes. Possible cases for using version-dependent master data are simulations of different shift models or lot sizes. By design it is possible to use different versions of any model for simulation purposes, and most of the SAP APO™ components provide simulation options. Exceptions are ATP and TP/VS, since the associated tasks are very close to execution. The precondition for a simulation is copying the active version (i.e., the version which contains the operative data and is connected with SAP ERP™) to a simulation version. However, the data volume is usually huge and the increase in hardware memory requirement with the number of simulation versions is almost linear. Though simulations are quite often desired (*cf. desirable features 2.3.1.o, 2.3.3.1.u, 2.3.3.2.i, 2.3.4.2.d, 2.3.5.1.c, 2.3.5.2.i, and 2.3.6.2.j*), this prevents many companies from using simulations.

Some companies use the whole range of modules in SAP APO™, other companies apply only one or a few modules – there is hardly a combinatorial possibility for which no implementations exist.

4.2.1 Demand Planning

The purpose of Demand Planning (DP) is to forecast customer demand before customers actually place their orders. The results of Demand Planning are forecasts per product and location. In the case of make-to-stock the forecast is used to determine the production quantities of the finished products; in other cases the forecast might only be used for the production of assembly groups or the external procurement of components. Demand Planning is applied for tactical planning with a typical planning horizon of 12–18 months. In some cases – especially in the consumer product industry – demand planning is also used for operative planning to adjust the short-term forecast to recent sales numbers. Fig. 4.3 points out that the nature of DP is not collaborative. Though the object of DP is the customer's demand, planning is performed by the company that owns the SAP APO™ system. This is referred to as the own company, and unless mentioned otherwise this will be the point of view for the process descriptions.

Demand Planning is usually performed at an aggregated level – i.e., for product groups, customer groups, and regions. One reason for this is that the accuracy of demand planning is generally better when aggregated data is used. Another reason is that in most cases it is not possible to check and plan the forecast interactively for each product on a regular basis. The planning levels are defined ad libitum (*cf. desirable feature 2.3.3.1.k*). By default, a consistent planning method is used: even though demand planning is performed on an aggregated level, the data is disaggregated and saved at the lowest (most detailed) level. The standard disaggregation method is pro rata, i.e., in proportion to the previous values of the members (for instance, if member A has a demand value of 10 and member B of 20, and the aggregated value is increased from 30 to 45, the disaggregation pro rata will result in values of 15 for A and 30 for B).

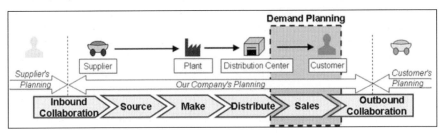

Fig. 4.3 Scope of Demand Planning within SCM

The main input for DP is the demand history. The procedure to determine the forecast varies from a simple interactive planning based on the planner's experience to a complex scenario including statistical forecasting and multiple calculations and comparisons. Fig. 4.4 shows an example of a demand planning process in which the planning department creates the forecast based on three different inputs: the information they receive from the sales department, the forecast history (e.g., shifted by 1 year), and the statistical forecast. Additionally, the impact of promotions is considered (process step "promotion planning") and some simple calculations are performed by macros – e.g., a proposal for the forecast is calculated as a weighted average of the three inputs, the forecast is cumulated in order to check the conformance with the annual targets, and alerts are calculated for huge deviations (process step "background calculation steps"). Based on this information the planner determines the final forecast and releases it – the result is a forecast at product and location level, which becomes relevant for procurement, production, and distribution planning.

Life cycle planning in SAP APO™ considers the impact of the product life cycle – both for products at the end of their life cycle and for newly introduced products. For a newly introduced product the forecast is calculated using the history of similar products and is adjusted by a phase-in profile which dampens the forecast values in order to achieve a ramp-up slope (*cf. desirable feature 2.3.3.1.c*).

It is not necessary to execute all these process steps, and even the sequence of the steps may vary. Similarly the assignment of the tasks to the organization units might differ – sometimes the entire demand planning is performed by sales only or by production only. Demand planning is often performed in monthly buckets, but in some industries – e.g., consumer products – the granularity is finer for the near future, and therefore mixed buckets are used to plan with weekly or in some cases even daily buckets in the near future.

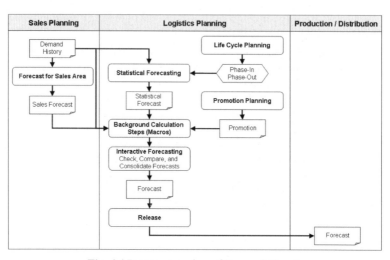

Fig. 4.4 Process overview of Demand Planning

(A) Statistical Forecasting

The accuracy of the demand history is one of the key factors in forecasting. Potential causes of inaccuracy are incomplete demand history and mismatches in the unit of measure. Another issue is that of demand irregularities (e.g., because of accepted tenders), which have to be cleaned interactively in order to provide a representative demand history for statistical forecasting.

There are two major groups of statistical forecasting methods. One group consists of univariate methods, where sales figures are forecast from their past values, and the other group, of causal models, where the sales figures are forecast depending on the history of other figures (also referred to as "causal factors"). For the latter the most commonly used method is multiple linear regression (MLR) (*cf. desirable feature 2.3.3.1.i*). The quality of the statistical forecast depends on the successful approximation of inherent regularities, and analysis and prediction of these regularities become better as the number of data elements increases. At the determination of the forecast level the trade-off between increasing the amount of data by using a higher aggregation level (e.g., product group instead of product) and losing information through inadequate disaggregation has to be considered. A common scenario is a forecast carried out on the more detailed level (e.g., product level) for the short-term horizon, when the accuracy of the operative data has priority, with a switch to forecasting on an aggregated level for longer horizons. The logical precondition for forecasting at the aggregated level is that the history of each aggregated product is similar. A further issue is the selection of the data basis. A typical question is whether to use third-party sales or a mix of third-party and intra-company sales for forecasting. To avoid distortion of the inherent regularities of the sales data by local inventory policies and lot sizes, whenever possible only third-party sales should be used as data history.

Another crucial factor is the application of the appropriate forecast model, which is able to model these regularities. The selection of the forecast model based on an analysis of the data history has a significant impact on the accuracy of the statistical forecast.

Composite forecasting allows the calculation of several forecasts with different models and parameters and either selects the one with the lowest retrospective forecast error or a weighted combination of the forecast (Dickersbach 2006). The retrospective forecast error is determined by the difference between the historical data and the forecast model applied for the past. The forecast models are assigned interactively or selected automatically, based on error measures. It is possible to perform an automatic outlier correction for the demand history during forecasting. Automatic outlier correction is either based on the ex-post forecast (which already contains the impact of the outlier, so that the results might differ from expectations) or based on the median forecast (Dickersbach 2006). Forecasting may also be used in combination with Collaborative Demand Planning, as described in Section 4.2.3.A (*cf. desirable feature 2.3.3.1.g*).

(B) Macros

Macros allow processing of the huge amount of planning data with customer-defined calculations. Examples for macros are calculations of differences and mean values of different forecasts. With macros it is also possible to access master data properties – e.g., lot sizes – and to change the format of cells depending on rules in order to visualize remarkable values, e.g., a deviation from last year's sales by over 50%. For more sophisticated calculations it is possible to apply user function or user exit macros. Another use of macros is the creation of alerts, e.g., deviations or threshold trespasses. In this way the alerts can be defined very flexibly. The handling of alerts is described in Section 4.6.B.

(C) Interactive Planning

The creation of an adequate forecast is the planner's responsibility, and therefore SAP APO™ offers a highly configurable planning book for a clearly arranged presentation of the data. For DP one unit of measure is used in order to keep the aggregation and disaggregation of data semantically correct. Therefore, all planned products must contain this unit of measure either as base unit or as an alternative unit of measure. It is, however, possible to switch between different units of measure (if the units of measure are maintained in the product master). When planning on different levels it is possible to lock values using the functionality for fixing. As an example, the planner expects sales for portable MP3 players (as a product group) of 12,000 pieces, but for a certain player – e.g., 30 GB black – he is pretty sure that 4,000 pieces will be sold. Fixing is required in order to protect this value of 4,000 pieces against changes resulting from a change at product group level. If the forecast for portable MP3 players is updated to 10,000 pieces, the forecast for the 30 GB black player remains 4,000 pieces, and the forecast for the other members of the product group is reduced by a larger than average number. The other way round it is possible to fix the aggregated value – e.g., 12,000 pieces for portable MP3 players – with the result that a change in the forecast for one member of the product group is counterbalanced by the other members.

To enable offline planning (e.g., for sales representatives) also, it is possible to download the data as displayed in the planning book to Excel, change the data, and upload it into DP.

(D) Promotion Planning

Promotions are mainly used in consumer products industries to create or hold customer awareness for a product or a brand. The effect of the promotion might be an increase in demand or just a shift of the demand profile in time. Promotion planning functionality of SAP APO™ supports planning the impact of promotions on demand. Promotions are modeled as an absolute or a relative increase in sales (*cf. desirable feature 2.3.3.1.h*). Unlike life cycle planning, promotion planning is not integrated into the statistical forecasting process but applied independently of the statistical forecast. It is possible to plan promotions on different levels without mixing the data. The cannibalization effect of a promotion (i.e. the decrease in the demand for a related product caused by a promotion) is modeled by cannibalization groups. The cannibalization is planned within the same time bucket and does not allow modeling of a post-promotional dip.

(E) Forecast Release
Only the released forecast is relevant for the subsequent planning processes. Though demand planning can be performed at any level, the forecast release has to be at location and product level. If the location is not modeled at planning level, the location split table defines how to release the demand to the locations. By analogy with the location split, it is also possible to define a product split. In both cases flexibility for the assignment of the forecast is lost. The time granularity for demand planning is often coarser than for distribution or production planning. Therefore the forecast can be disaggregated during release from a monthly time bucket to daily forecasts, for example. The forecast is either released within SAP APO™ to the modules SNP and PP/DS or transferred directly to SAP ERP™ as a planned independent requirement (e.g., if SNP and PP/DS are not used).

(F) Forecast Netting (Forecast Consumption)
In a pure make-to-stock scenario distribution, production and procurement planning are based entirely on the forecast and not on sales orders. In this case the forecast needs to be adjusted to deviations in the customer demand. Alternatively, it is possible to net the forecast by sales orders immediately (also referred as "forecast consumption") so that the sum of sales orders and unconsumed forecast becomes relevant for the subsequent planning steps. Parameters of the consumption are the time difference between forecast and sales order and the direction (forward, backward, or both) (*cf. desirable feature 2.3.3.1.q*).

The forecast is netted at the level of product and location. It is possible to refine this by additional levels, such as customer or customer group (Dickersbach 2005).

(G) Demand Planning for Assembly Groups
In some cases the finished product is only produced if an order already exists. To keep the order fulfillment lead time short, it is nevertheless desirable to produce assembly groups in advance. In these cases demand planning is either performed at assembly group level (instead of finished product level) (*cf. desirable feature 2.3.5.2.g*) or the last production step for the final product – its assembly – is only planned, and executed only if an order is placed (*cf. desirable feature 2.3.5.2.f*).

(H) Characteristics-Based Forecasting
In most cases when configured products are used it is not sufficient to plan the product, because the configuration often determines the dependent demand for the assembly groups and components via object dependencies. With the functionality for Characteristics-Based Forecasting (CBF), demand planning is performed for the configuration as it is for another planning level (e.g., customer or product group). However, it is only possible to plan the configuration of the finished product (and not of the assembly groups).

(I) Dependent Demand for Complex Products or Active Ingredients
In the consumer product industry sets or kits are often planned – especially for promotions – which contain several finished products or a multiple of one finished product. Information about the dependent demand is required to gain an overview of the total demand and to plan cannibalization effects. Another case for the use of dependent demand in demand planning is given if a key component limits the supply quantity. If many products contain this key component – probably even in different quantities, as in the case of the active ingredients in the pharmaceutical industry – a rough feasibility check by rule of thumb is not possible. For this purpose it is possible to use BOM information in DP and display the dependent demand within the planning book. An alternative procedure is to upload the dependent demand from distribution or production planning.

(J) Forecast After Constraints
A feasibility check for the demand plan is helpful to allow the earliest possible reactions to potential capacity constraints. In the case of infeasibility, decisions for capacity expansion, allocation or focusing on key markets can be triggered in advance. Whether a forecast is feasible or not is determined in SNP and/or PP/DS – if the inventory and the planned receipts cover the forecast, the forecast is feasible. For the feasibility check the inventory and the planned receipts are transferred to DP and compared with the demand. A suitable way to interpret these figures is by cumulating the demands and the supplies and checking whether the cumulated supply falls below the cumulated demand (*cf. desirable feature 2.3.3.1.l*).

4.2.2   Supply Network Planning

The purpose of Supply Network Planning (SNP) is tactical and/or operative cross-location planning, including sourcing decisions. The focus of SNP is the internal supply chain consisting of a distribution network and one or more plants (*cf. desirable feature 2.3.5.2.a*). SNP is designed for the internal supply network planning of make-to-stock products. Therefore it is not suited for products with so-called requirement strategies, such as "make-to-order" or "planning without final assembly." The same applies to configurable products. Fig. 4.5 visualizes the scope of SNP within SCM.

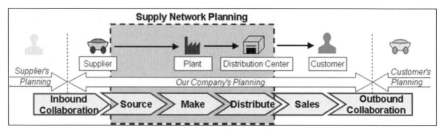

Fig. 4.5 Scope of Supply Network Planning within SCM

Tasks in Supply Network Planning are:

- Safety stock planning determines appropriate safety stock levels for achieving the desired service level.
- Distribution planning calculates the net requirements at each location and propagates the net demand to the source location taking transportation times and lot sizes into account. In the case of multiple sourcing within the internal supply chain (e.g., replenishment of a distribution center from two different plants) the source is selected.
- Rough-cut production planning creates planned orders with a lower level of detail to check the capacity requirements and the feasibility of the supply chain demand and triggers the procurement of key components with long lead time. Within the time bucket no sequence exists for the planned production orders.
- Integrated distribution and production planning allows improvements of the service level and resource utilization in the case of multiple sourcing alternatives (cf. desirable feature 2.3.5.2.c).
- Replenishment plans the operative distribution of the available goods and decides how much to deliver to whom in the case of shortage.

Distribution and rough-cut production planning are performed for a horizon that covers the short and medium term, and probably even the long term. Fig. 46 provides an overview of the different tasks.

The objective of distribution planning is to determine the net requirements at each location of the supply network – including the safety stock – and to plan stock transfers with due consideration for the transportation duration. For the scheduling of stock transfer orders the duration for goods issue, transport and goods receipt is taken into account with the corresponding calendars. Distribution planning of the internal supply network gains importance, since many companies change their processes from non-coordinated local inventory management to a global inventory

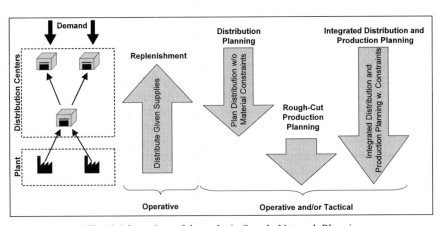

Fig. 4.6 Overview of the tasks in Supply Network Planning

management in order to reduce inventories. By concentrating safety stocks from several local Distribution Centers (DC) to a central DC and changing the responsibility for the inventories at the local warehouses combined with service level agreements, significant stock reductions are achieved. SNP is primarily used in the consumer products, high tech, and chemical industries for make-to-stock products in an extensive supply network, but is becoming more important for other areas as well, for example because of mergers leading to extensions of internal supply chains. Of all SCM modules, SNP also has the largest potential impact on the organization, since it offers options for coordinated and integrated production and distribution planning.

In addition to the lead time for replenishment the following constraints might be considered in distribution planning with SNP:

- Transport capacity as a restriction of the transport volumes between locations.
- Especially in the consumer product industry, where a high volume of goods is dealt with, storage capacity might become a constraint (cf. *desirable feature 2.3.6.3.c*). A restriction in the SNP model is that only one storage resource per location is possible. This might be a problem if different qualities of storage capacities are used at one and the same location, e.g., for frozen and for non-frozen products. The storage resource cannot be used to model storage constraints in production (e.g., limited buffer capacity for bulk products before packaging).
- Each goods receipt and each goods issue consumes capacity. This handling capacity is represented by handling resources, which are assigned to the location (one for goods receipt and one for goods issue).

Another task for distribution planning is deciding from which source the demands are covered (if the supply chain is organized as a multi-sourcing network). Sourcing according to priorities represents a clear preference for particular sources, and any use of other sources is than an exception. Quota arrangements are used for sourcing from multiple locations on a regular basis. The SNP Optimizer, however, ignores priorities and quota arrangements and proposes the sources based on the total supply chain costs.

(A) Safety Stock
Safety stock is used to buffer for uncertainty on the demand and on the supply side. On the one hand it is desirable to keep the stock levels as low as possible, and many SCM projects start with the goal of reducing the stock levels within the supply chain. On the other hand, stock is required in order to meet the target service levels toward the customer – either at finished product level or for assembly groups. For distribution and sourcing, safety stocks are required to compensate the uncertainties of supply. The amount of safety stock required decreases with the lead time. Especially in complex internal supply networks, the transparency of the distribution may itself lead to a reduction of the safety stock: for instance, if the subsidiary of a consumer products company is able to monitor the planned distribution quantity and the stock in transit, it is likely to reduce its buffers against uncertainty on the supply side. Production safety stock is required in order to decouple the production processes.

In SAP APO™ safety stock is either defined interactively as an absolute quantity or as a target number of day's supply in the product master data. In cases of seasonal or irregular demand the safety stock cannot be defined as a fixed parameter, but as dependent on time. Using advanced safety stock methods it is also possible to calculate the safety stock levels based on a target service level, taking account of the forecast error (from DP or from the product master). However, these functions are not designed for calculating the safety stock of external supply chain partners as well (*cf. desirable feature 2.3.3.1.r*).

(B) Supply Network Planning Alternatives
There are various ways of tackling supply network planning, and different aspects have to be considered:

- Supply chain structure: Does multiple sourcing exist within the supply chain (on a regular basis, not as an exception)?
- Business requirement: Is it necessary to have a feasible distribution plan – i.e., to check in advance whether a demand can be covered in the distribution network?
- Organization: Is there a dedicated responsibility for joint distribution and production planning or are the responsibilities disconnected?

In the case of single sourcing there is usually not much potential for optimization of the entire supply chain (there might be potential for the optimization of the production). If there is no need for a feasible distribution plan either – e.g., because the demand is usually covered and buffered by safety stock against shortages, as is often the case for consumer products – SNP is much simpler than in a constrained multi-sourcing supply network, where a feasible distribution plan is required for product allocations. When traditional logistics concepts are applied, distribution planning and production planning are carried out completely independently of each other. Although the approach of SCM implies no separation of planning according to functions, in many cases, especially in that of single sourcing, a hierarchical step-by-step approach – first distribution planning and then production planning – is sufficient. According to different requirements, SAP APO™ offers three different methods for SNP:

- The SNP Heuristic,
- SNP Optimization, and
- Capable-to-Match (CTM).

Fig. 4.7 shows an example of the process flow of supply chain planning encompassing both distribution planning and rough-cut production planning. The results of distribution planning are planned stock transfers. Deployment adjusts the stock transfers to the available supply and Transport Load Building (TLB) groups or splits the stock transfers according to the capacity of the means of transport before execution.

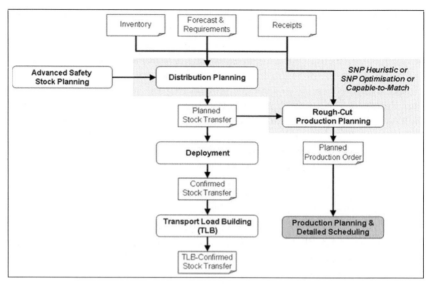

Fig. 4.7 Process overview of Supply Network Planning

(C) SNP Planning Periods and Planning Book

SNP uses time periods (also called "time buckets") for planning, and the smallest granularity is a day. Other time periods, such as weeks and months (or a mixture of these), are also possible. The purpose of planning with time buckets instead of planning individual orders is to aggregate planning (*cf. desirable feature 2.3.3.1.k*). The planning situation in the supply network is displayed in the SNP planning book. Quantities and capacities are displayed in separate data views. The quantity view relates to location products, while in the capacity view the resource is usually the relevant characteristic. Capacity consumption is displayed for production, transport, storage, and handling resources (*cf. desirable feature 2.3.6.3.d*). The quantity view contains the projected stock and information about supply shortages.

(D) SNP Heuristic

The SNP Heuristic is the preferred tool for supply network planning without constraints. Its advantage is its simplicity: It calculates planned stock transfers to cover the net requirement taking the safety stocks, the lot sizes and the transport duration into account – but not the capacities. The result of the SNP Heuristic is easy to understand, but the distribution plan might not be feasible. In the worst case it may result in lateness due to lead times, but no shortage can occur. In other words: The demand will always be balanced by supplies, though not necessarily on time.

The capacity requirements for production, transport, storage, and handling are displayed in the capacity view of the planning book. Capacity overloads are leveled in a separate process step, but no dependencies between the orders or operations and no component availability are considered. Overloads are distributed to other time buckets, backward, forward, or both, with options to prioritize according to the order size or the product priority. It is also possible to apply capacity leveling using the functionality of the SNP Optimizer – with the difference that the costs for the SNP Optimizer are calculated automatically according to the settings in the capacity leveling profile. The advantage is that no costs have to be maintained.

(E) SNP Optimizer
The SNP Optimizer plans the entire internal supply network – procurement, production, and distribution – at minimal costs by modeling the supply network as a system of linear (in-)equations. The best solution is determined by linear programming or mixed integer programming. The goal of the SNP Optimizer is the creation of a plan at minimal cost, where turnover and capital are partially modeled using penalty costs. Structural modification of the supply network – e.g., the shutdown of a distribution center – is not in the scope of the SNP Optimizer.

The supply chain costs for the SNP Optimizer contain costs for production, procurement, transport, storage, and handling; penalties are costs for lateness, non-delivery, and safety stock violation. These costs are expressed as relative numbers ("APO dollars"), and they are not transferred from SAP ERP™ but have to be maintained in SAP APO™. By setting these costs it is possible to model such decisions as

- extending production capacity in a plant or procure from a different plant considering increased production and/or transport costs (*cf. desirable feature 2.3.5.3.e*),
- extending production capacity in a plant or procure externally,
- switching to a more expensive transport method to reduce the transportation duration (e.g., from truck to plane) (*cf. desirable feature 2.3.6.2.f*),
- switching to another source because of transport capacity constraints, and
- producing and shipping just in time to minimize storage costs.

The SNP Optimizer is very well suited to the planning of a complex supply network with multiple decisions on sourcing and trade-offs. For a detailed treatment of APO's optimization capabilities we refer to Kallrath and Maindl (2006).

Probably the most critical issue in the application of the SNP Optimizer is the appropriate cost determination. If the costs are not defined in reasonable relations to each other, rather unexpected results may be attained, e.g., no production at all because supply chain costs exceed the penalties for non-delivery or permanent transport because storage costs are higher than the additional transport costs. The SNP Optimizer often reacts quite sensitively to inappropriate settings of the costs.

Constraints for optimization are the demands, the capacities, the material availability, and the fixed horizons for production and stock transfer. The optimizer profile controls whether capacities (for production, transport, storage and/or handling resources) are considered as finite or infinite, whether component availability is a constraint, and the priority of demand categories (e.g., forecast and safety stock).

The optimization problem becomes significantly more complex if integer variables are necessary to model the decision situation adequately. Some cases in which a discrete model formulation is required are

- lot size dependent cost, e.g., decreasing costs due to less set-up, or increasing costs if agreed procurement quantities are exceeded,
- lot size independent capacity requirements, e.g., for set-up,
- technical or economic restrictions requiring a fixed lot size, lot size rounding, or a minimum lot size,
- capacity extension, and
- prioritization of the production in case of shortages.

Each case for discretization is controlled by a separate parameter, and each parameter and each time bucket for which discretization is used will complicate the planning problem. The performance of the optimization depends, in particular, on the number of locations, products, and buckets and on whether and to what extent discretization is used. Since setting up an appropriate optimization scenario is a rather complicated task, SAP offers a special consulting service for optimization.

(F) Capable-to-Match

CTM performs an iterative approach where demand elements are prioritized, supply elements are categorized, and demands are matched with the supplies. CTM planning leads to a highest priority – first serve approach. CTM is therefore best suited to situations in which demand priorities exist – examples are demand types (e.g., special orders), customers, or products (e.g., highest profit) (*cf. desirable features 2.3.3.2.e, 2.3.5.2.h, and 2.3.6.2.c*). CTM is not an optimization method but a heuristic, and there will be no re-planning of orders which are already assigned to other demands. The solution will therefore not necessarily be optimal.

An additional feature of CTM is the option to push supplies from source to target locations even when there is no demand for them – e.g., in order to free up storage capacity at the source (*cf. desirable feature 2.3.6.2.c*).

CTM is able to use either the less detailed master data for the SNP module or the more detailed master data for the PP/DS module. Using the SNP master data, the planned orders are split according to the daily free time buckets if the order size exceeds the bucket capacity. If set-up is modeled as fixed capacity requirement, the set-up is considered for each time bucket (see Section 4.2.2.C). Fixed or minimum lot sizes have to be used with great care, because in this case the planned orders are no longer split. This might lead to very low utilization and – if the capacity requirements exceed the daily capacity – to shortages due to unplanned orders. Fig. 4.8 shows an example of a demand that can be covered on time if the planned orders are split to utilize the remaining free capacity of the first two periods (case A). Because in case B a fixed lot size that requires the capacity of a whole period is produced, the demand is covered late. In unfavorable circumstances the fixed lot size might even lead to a capacity requirement in excess of the capacity available, with the consequence that the demand will not be covered at all as shown as case C.

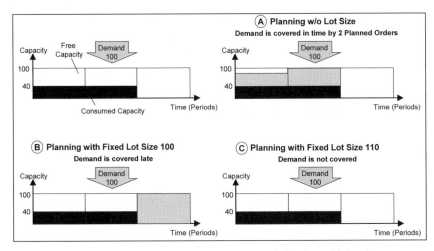

Fig. 4.8 CTM with SNP master data: Impact of planning with lot sizes

When CTM is used for production planning with PP/DS master data, some restrictions exist compared to PP/DS planning. For example: sequence-dependent set-up cannot be modeled, and the operations have to be in a linear sequence. Since CTM performs a one-step finite planning it might not lead to a sufficiently good schedule. Therefore, a subsequent scheduling step in PP/DS should be considered.

CTM also offers an option for supply distribution which pushes supplies to the target locations even when there is no demand for them. This might be required if the production plant does not keep the stock itself but immediately passes it on to the DC. All free supplies are pushed to the target locations.

(G) Comparison of the SNP Solution Alternatives
Table 4.1 lists the main features of the three SNP solution alternatives.

Table 4.1 Features of the SNP solution alternatives

	SNP heuristic	SNP optimizer	CTM
Feasible distribution plan	No	Yes	Yes
Feasible production plan	In simple cases after capacity leveling	Yes	Yes
Dynamic sourcing	No	Yes	Yes
Supply allocation in case of shortage	n.a.	By cost	By priority
Production capacity	Infinite	Finite	Finite
Transport capacity	Infinite	Finite	Finite
Storage capacity	Infinite	Finite	Ignored
Handling capacity	Infinite	Finite	Ignored

By dynamic sourcing we understand the selection of sources for supply based on capacity and lead time constraints. Infinite capacity planning creates capacity requirements, whereas no capacity requirements are created when capacity is disregarded.

(H) Replenishment with APO
There are two processes within replenishment: Deployment and transport load building. Deployment calculates the fair share of quantities to the requesting parties in cases of shortage or surplus. The only constraints are the quantities available. Transport load building is one step closer to execution and focuses on the creation of truck loads, where the task is to assign the planned stock transports to the available transport means by taking their capacity restrictions into account.

For the replenishment process the main choices available are whether to use the deployment heuristic or the deployment optimizer to determine the quantities, plus the option of using or not using the Transport Load Builder (TLB) to take transport resource restrictions into account and arrange the stock transfer orders accordingly. The four alternatives are visualized in Fig. 4.9.

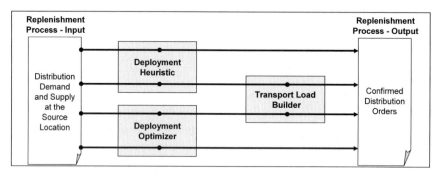

Fig. 4.9 Alternative processes for Replenishment

(I) Deployment
If the supply chain behaves exactly as planned – i.e., neither changes in the demand nor unpredicted deviations of the supply occur – deployment is not necessary. However, there will usually be differences between planned and available quantities when it comes to execution. If the demand exceeds the supply, it is necessary to decide which demands will be covered from which locations and to what extent (*cf. desirable feature 2.3.6.2.d*). The supply chain structure determines the complexity of these decisions. In a hierarchical, single sourcing structure a simple fair share rule is sufficient. In a multi-sourcing structure the decisions become more complex and might require deployment optimization.

Deployment converts planned stock transfers into confirmed stock transfers according to the available supplies, the demands, the deployment strategy, and the fair share rule. The Deployment Heuristic is performed location by location for all source locations. If the Deployment Heuristic is used immediately after finite distribution planning (by SNP Optimization or CTM), a fair share situation does not exist because planned stock transfers were created only for the available quantity and not for the original demand (see Sections 4.2.2.E and 4.2.2.F). In this case the Deployment Heuristic only confirms the planned distribution orders.

The point in time of the stock transfer – whether stock is rather kept at the source or at the target location – is defined by the deployment settings. Examples of options are a confirmation of the distribution orders according to the requirement date of the planned stock transfer orders at the source location (i.e., without rescheduling), scheduling of the confirmed stock transfer orders as early as possible, and shipping all receipts within a defined horizon to the target locations according to the outbound quota, regardless of the actual requirements. If no push is allowed for deployment, because certain locations have to receive supplies just in time, this is defined in the location master data for the target location.

During deployment the requirements are processed in the order of their due dates (in the granularity of the assigned time buckets), so that shortages always affect the requirements farther in the future. For demands that are due in the same time bucket, the fair share rule defines which demands are fulfilled and to what extent. Fair share rules are "percentage distribution by demands," "percentage fulfillment of target," "percentage division by quota arrangement," and "division by priorities" (*cf. desirable features 2.3.6.2.d and 2.3.6.3.e*). These rules are explained in more detail by Dickersbach (2006).

The Deployment Heuristic is based on the stock transfers that were previously planned by the SNP Heuristic (or another function). In order to be more up-to-date, the Real-time Deployment performs the SNP Heuristic to determine the current requirements before the deployment calculation.

If a plant or a central DC is not exclusively dedicated to supplying other DCs but also keeps inventory for delivering customers directly, the question of the prioritization of sales orders versus distribution requirements arises. A distinct prioritization is modeled by the definition of the order categories for the available quantities. It is possible to define a fair share between sales orders, forecasts, and distribution requirements.

For multi-sourcing supply chain structures the Deployment Optimizer determines the stock transfers based on the current demand situation in the network, taking sourcing alternatives and balancing or distributing shortages into account. This is a valuable advantage, especially if the SNP Heuristic is used for distribution planning.

(J) Transport Load Building

The Transport Load Builder is a short-term planning tool for combining confirmed distribution orders to truckloads or other transport units according to the capacity constraints (*cf. desirable feature 2.3.6.3.g*). Inputs for TLB are the confirmed stock transfer orders from deployment, and orders for different products are combined to TLB-confirmed stock transfer orders. TLB-confirmed stock transfer orders are also created and changed interactively.

The algorithm for TLB offers options for upsizing or downsizing the quantity of the stock transfer orders and for pulling future orders forward to combine them with due orders in order to increase the utilization of the means of transport. Additional features of the new TLB algorithm are parameters to control the mixture of the products in shipments, as shown in Fig. 4.10. "Straight loading" combines products of the same loading group and is more efficient for loading and unloading, while "load balancing" distributes products evenly to different truckloads and reduces the risk of stock-outs if one shipment fails.

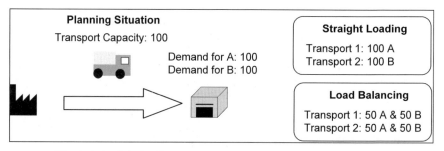

Fig. 4.10 Straight loading and load balancing in Transport Load Building

Capacity constraints are defined in the TLB profile as the minimum of the number of units allowed by the maximum volume, the maximum weight, and the maximum number of pallets. There is also a lower limit to inhibit uneconomic transport orders. The pallet restrictions are maintained in the TLB profile as pallet floor spots (the number of pallets that can be placed beside each other). The capacity consumption of the pallet floor spots takes the stacking factor of the product into account. This capacity calculation assumes that pallets with different products are stacked together, which is not the case in all businesses – e.g., there are restrictions on the combinations of products for transport purposes (e.g., transport of gas cylinders containing oxygen and hydrogen together is forbidden because of the danger of explosion). Incompatibilities of this kind cannot be modeled with TLB. The capacity of the transport resource is disregarded during the TLB run.

(K) Rough-Cut Production Planning
In SNP, scheduling and capacity planning are independent steps that each use a different set of data (as in SAP ERP™). While the duration of an order is always a multiple of a day and independent of the order quantity, the capacity consumption is usually variable (i.e., proportional to the order quantity) – but a fixed capacity consumption can also be handled.

An implication of fixed order durations in SNP is that a planned order takes at least 1 day. Since SNP is intended for rough-cut production planning, fixed order durations are usually sufficient. It is assumed that the order quantities do not vary widely enough to cause considerable prolongations of their production lead times. If this is not true and the prolonged lead times are relevant for planning, one approach is to use different master data for production, with different durations for the distinct lot size intervals. Because of the fixed durations SNP is not suitable for detailed scheduling.

(L) Coordination with PP/DS and SAP ERP™
The move from SNP to PP/DS contains a change from bucket-oriented to time-continuous planning and involves a completely different set of production master data. There are basically three options for the move to PP/DS, and each one has its own disadvantages:

- Conversion of the SNP planned orders to PP/DS implies that no production planning is done in PP/DS. Therefore it is not possible to react to short-term changes in PP/DS.
- Production planning in PP/DS deletes the SNP planned orders and creates PP/DS planned orders to cover the net requirements. This has the disadvantage that leveling or prioritization by SNP is overruled.
- SNP planning is performed in an inactive version and the SNP planned orders are transferred to PP/DS as planned independent requirements with the disadvantages of losing transparency between SNP and PP/DS and the effort involved in data replication.
- If no PP/DS is used in SAP APO™, the SNP orders are transferred directly to SAP ERP™.

(M) Sourcing of Forecast
"Sourcing of forecast" describes a process where demand planning is performed at customer level, and the customer has several ship-to locations which are supplied from different DCs of the internal supply network (multi-sourcing is possible). The forecast is transferred to the customer location, and the decision from which DC to source the demand is made by the SNP Optimizer, involving different forecast re-calculation steps in DP, allocations (see Section 4.2.5.A), and rules-based ATP (see Section 4.2.5.B). The benefit of an optimized allocation of the forecast demand to the DC has to be balanced against the significant increase in the complexity of the planning process.

(N) Aggregated Planning
Aggregated planning is performed in SNP using product hierarchies with pro-portional factors (*cf. desirable feature 2.3.3.1.k*). The master data representing the aggregates – the (dummy) product for aggregated planning, the associated (dummy) master data for production, and the (dummy) resource – have to be created inter-actively, and it is up to the planner to maintain the parameters in a meaningful way. The following steps are performed in aggregated planning:

- Mapping of the detailed demands and supplies in order to ensure that a receipt for product A does not cover a demand for product B after aggre-gation.
- Aggregation of the demand and the supplies; an aggregated planned order is created for existing detailed planned orders.
- Distribution and rough-cut production planning at aggregated level.
- Disaggregation of the aggregated orders creates detailed orders.

The default disaggregation is pro rata. In periods without detailed orders this leads to an even distribution, which is not always desired. Therefore it is also possible to disaggregate according to the proportional factors of the product hierarchy or to apply customer-specific logic for disaggregation.

A different kind of aggregated planning is planning with aggregated resources (in contrast to a dummy resource as in the previous case). This might be interesting if multiple resource alternatives exist and assignment to an individual resource is not yet required. In this case a resource hierarchy and a hierarchy for the production master data are required with an aggregated resource and associated master data (cf. Section 4.7.3). The capacity consumption is aggregated to the aggregated resource and planning – including capacity leveling – is performed with the aggregated production master data. In a subsequent disaggregation step, planned orders are created with detailed resources. It is impossible to combine the two alternatives – planning is performed either with aggregated products or with aggregated resources.

4.2.3   Supply Chain Collaboration

From a planning point of view, we understand by supply chain collaboration the planning with mainly external partners. However, only the collaboration with the partners immediately adjacent in the supply chain – the customer and the supplier – is considered and not that with the customer's customers or the supplier's suppliers (Fig. 4.11).

For customer collaboration SAP APO™ offers processes for collaborative demand planning and for Vendor Managed Inventory (which cover tasks for demand and supply management and for execution of the CPFR® model, as described in Section 2.1.2.4) (*cf. desirable feature 2.3.3.1.e*). For supplier collaboration the demand information is shared via the Internet. Collaborative Management of Delivery Schedules (CMDS) enables a symmetrical collaboration between the customer's SAP APO™ and the supplier's SAP APO™.

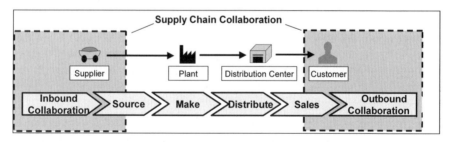

Fig. 4.11 Scope of supply chain collaboration within SCM

(A) Collaborative Demand Planning
Collaborative Demand Planning provides the option of interacting with internal or external customers in the demand planning process via the Internet (*cf. desirable features 2.3.3.1.a and 2.3.3.1.m*). Examples are the involvement of key customers

and the integration of subsidiaries with a less developed IT infrastructure. Possible scenarios are:

- Demand transmission, where the customers maintain their forecasts them-selves via the Internet in the SAP APO™ planning book of the supplier. This procedure has the disadvantage that it is not possible to load data – e.g., from an Excel sheet – into the planning book when accessing it via the Internet.
- Demand confirmation, where forecasting for the customers is done by the supplier and the customer just confirms or adjusts the forecast.
- Sales exception notification, where sales representatives report major deviations from the previously forecast values as soon as possible. This information is entered without any delay into SAP APO™ via the Internet and evaluated by the planner, who is probably supported by alerts (see Section 4.6.2).

Collaborative Demand Planning can also be used for the coordination of promotions (*cf. desirable feature 2.3.3.1.p*). From a technical point of view, the basic concept of Collaborative Demand Planning is to allow a user to log on to a "foreign" SAP APO™ system and to provide access to the planning book via the Internet. Compared with the access via SAP GUI the functionalities are limited.

(B) Vendor Managed Inventory

The general idea of VMI is the replenishment of the customer's warehouse within agreed stock levels or service levels based on the current stock level at the customer's site and the demand forecast for the customer. The VMI functionality in SAP APO™ relies on agreed stock levels (*cf. desirable feature 2.3.3.1.f*).

The forecast for the customer is performed by the vendor based on the customer's sales history. VMI planning denotes the distribution planning from the DC to the customer, which is done with any of the SNP distribution planning methods – usually with the SNP Heuristic – and results in a planned stock transfer from the supplying plant or DC. VMI planning is performed either as a part of SNP or as a separate step before SNP is run. Deployment and Transport Load Building create a TLB-confirmed stock transfer order, which is transferred to the vendor's SAP ERP™ system as a sales order. The sales order triggers the creation of a purchase order in the customer's SAP ERP™ system via Application Link Enabling (ALE). With the posting of the goods issue the stock in transit is created in SAP APO™ as a receipt at the customer location. The booking of the goods receipt at the customer creates a proof of delivery in the vendor's SAP ERP™ system via ALE, which reduces or deletes the stock in transit. Fig. 4.12 shows an overview of the VMI process.

(C) Collaborative Procurement

There are two options for collaborative procurement in SAP APO™ for providing the supplier with a preview of the planned requirements. The first option is the use of a scheduling agreement that is similar to SAP ERP™. Scheduling agreements are used if a certain product is procured frequently from a supplier; they are

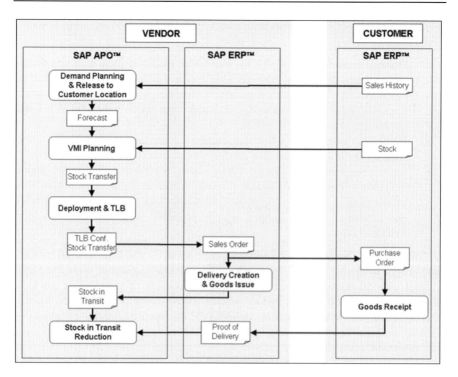

Fig. 4.12 Process overview of Vendor Managed Inventory

therefore especially popular in the automotive industry. In order to reduce the number of documents, only one scheduling agreement is used for multiple orders. The procurement proposals are called "schedule lines," the orders that are sent to the supplier are named "releases," and the planning result is called the "delivery schedule." The collaborative aspect of scheduling agreements is that a preview of the delivery schedule is also sent to the supplier as a "forecast release" (*cf. desirable feature 2.3.5.3.a*). In addition, it is possible to restrict the allowed deviations for the next delivery schedule per time horizon (both for the releases and for the forecast releases). The scheduling agreement has a target quantity, and the quantities ordered and received are reported against this target quantity.

Another option is to display the result of the production planning run in a planning book via the Internet (*cf. desirable feature 2.3.5.3.a)* with the same technology as used for Collaborative Demand Planning (see Section 4.2.3.A). For collaborative procurement via the Internet with scheduling agreements the supplier maintains confirmations in the planning book. The SNP Heuristic considers these confirmations in the next planning run.

(D) Collaborative Management of Delivery Schedules
CMDS is a process between partners with more or less equal rights and using scheduling agreements. The target industry for CMDS is the automotive industry and the key suppliers to it. One of the main advantages of this process is that not

only the customer demand is made available to the supplier, but also the supplier's capability is communicated to the customer by confirmations. Differing from other collaboration processes, in this way the supplier's answer is reflected in the customer system – in other words, the supplier can say "no" (*cf. desirable features 2.3.3.1.o and 2.3.5.3.a*). Fig. 4.13 shows the process overview of CMDS. The process steps performed automatically are underlaid with gray. All process steps are performed in SAP APO™, and the customer's and the supplier's systems communicate via EDI.

The customer creates the delivery schedule, consisting of binding "releases" and informative "forecast releases," and sends it to the supplier. The supplier uses the sales scheduling agreement in SAP APO™ and checks the delivery schedule for deviations from the previous delivery schedule. If the deviations exceed the agreed limits, the requests are corrected to the limits. These steps are performed by the admissibility check. In the next step the supplier performs its production planning and scheduling in order to cover the requirements. Based on the available stock and the planned receipts, backorder processing (see Section 4.2.5.G) is performed and the releases are confirmed – but not necessarily for the requested date with the requested quantity. The supplier monitors the status of the confirmations and is informed about deviations in the "sales scheduling agreement planning table." There it is also possible to override the confirmations or re-allocate the confirmation from another customer interactively. The confirmations are finally sent to the customer, and alerts notify the customer if there are deviations from the original request.

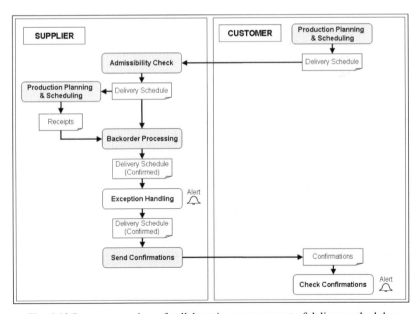

Fig. 4.13 Process overview of collaborative management of delivery schedules

4.2.4   Production Planning and Detailed Scheduling

The main objective of PP/DS is to determine a feasible production plan, including the planning of external procurement. The focus of PP/DS is operative planning, but it might also be extended to include tactical planning. Within the supply chain, the focus of PP/DS is the individual plant (Fig. 4.14). The capability of PP/DS for cross-plant planning is quite limited because it does not provide any cross-location view, it is not capable of dynamic sourcing based on capacity, and it does not offer deployment functionality.

PP/DS covers the process steps production planning, detailed scheduling and production control (Fig. 4.15).

Production planning creates planned receipts (planned orders for in-house production and purchase requisitions for external procurement) to cover the net requirements. After calculation of the net requirements, the alternatives for in-house production or external procurement are selected, and the procurement quantity is determined.

Detailed scheduling creates a feasible production plan (considering the available capacity and production constraints) by scheduling the orders. The result of detailed scheduling is a sequence of operations per resource. Criteria for the quality of the plan are (e.g.) resource utilization, effort involved in set-up, and meeting the due dates for the finished product.

Though production control is done in SAP ERP™, the current production situation is closely linked with PP/DS to take the present circumstances into account. The conversion of planned orders into production orders is triggered in SAP APO™ but the conversion itself is performed in SAP ERP™ (only in this way are the operation dates of the planned order kept). The release and the confirmation of the production order are performed in SAP ERP™; the order confirmation reduces its capacity requirement and updates its schedule in SAP APO™.

The horizon for detailed scheduling is usually shorter than the horizon for production planning. On the other hand, a feasible plan might be required for the medium term in order to detect shortages in advance. Detailed scheduling is performed on plant level. Switching a production order to a different plant (e.g., because of capacity restrictions) requires multiple interactive steps and is usually done in SNP.

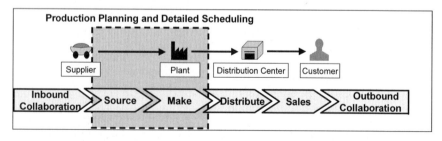

Fig. 4.14 Scope of Production Planning and Detailed Scheduling within SCM

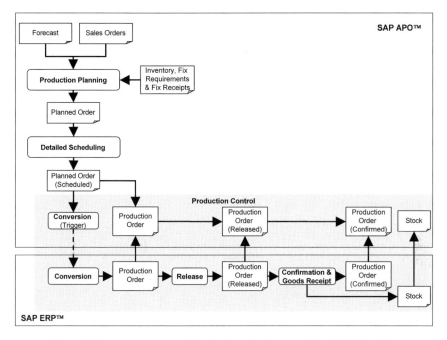

Fig. 4.15 Process overview of Production Planning and Detailed Scheduling

(A) Pegging
The orders are linked in PP/DS via pegging. It connects demands and supplies for the same product and the same location, and thus enables tracking of order networks from the customer demand to the purchase requisition (*cf. desirable feature 2.3.4.3.b*). In the case of dynamic pegging – which is the default – the pegging relationship changes with each new receipt or requirement because the pegging relationships are by definition not allowed to cross (Dickersbach 2006). It is also possible to fix the pegging relationship, e.g., for considering restrictions regarding batch pureness already during planning. The fixed pegging relationship is kept throughout the document changes for planned orders, purchase orders, and sales orders and can be considered in the ATP check. Pegging relationships may also be used for production constraints such as the maximum time between supply and demand (e.g., for molten metal) or for shelf life.

(B) Creation of Feasible Plans
The creation of feasible plans is usually performed in two steps: first production planning to cover the net requirement with planned orders or purchase requisitions and, as the second step, detailed scheduling in one or more scheduling steps for creating a feasible sequence for in-house production orders. Production planning is typically performed as an MRP run. Other methods of production planning in

PP/DS are the conversion of the SNP planned orders, production planning with CTM (see Section 4.2.2.F), and order creation triggered by the ATP check of the sales order, i.e. via Capable-to-Promise (CTP, see Section 4.2.5.C) or Multi-level ATP (see Section 4.2.5.D). Fig. 4.16 gives an overview of the different functions for production planning and detailed scheduling.

The planned orders that are created by the production planning applications are scheduled finitely or infinitely, but in either case detailed scheduling is required at a later stage. Creating a feasible plan in one step – i.e. the simultaneous creation and scheduling of planned orders – has proved rather unsuccessful in most cases, because the sequence determined in production planning is not suitable as schedule in most cases. Common problems were low resource utilization due to scattered resource loading, "loser products" because orders for products which are planned last have a high probability of becoming delayed in the case of capacity shortages, and high set-up efforts.

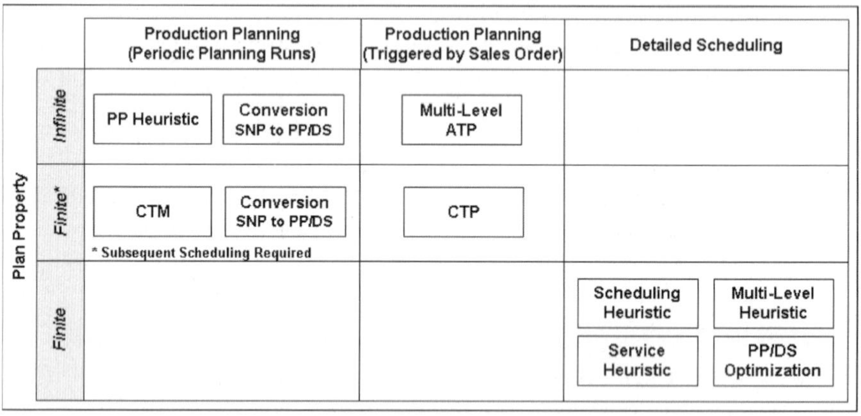

Plan Property		Production Planning (Periodic Planning Runs)		Production Planning (Triggered by Sales Order)	Detailed Scheduling	
	Infinite	PP Heuristic	Conversion SNP to PP/DS	Multi-Level ATP		
	Finite*	CTM	Conversion SNP to PP/DS	CTP		
		* Subsequent Scheduling Required				
	Finite				Scheduling Heuristic	Multi-Level Heuristic
					Service Heuristic	PP/DS Optimization

Fig. 4.16 Functions for Production Planning and Detailed Scheduling

(C) Heuristics for Detailed Scheduling
For detailed scheduling a variety of heuristics exists which is structured into service heuristics, scheduling heuristics, and multi-level heuristics. In addition to the standard heuristics it is possible to create customer-specific heuristics with moderate effort using the heuristic framework.

The purpose of service heuristics is rescheduling the orders and operations (which have already been scheduled without considering capacity constraints during production planning, see previous section) into the sequence of the material flow. An example is shown in Fig. 4.17, where the order for the finished product A had been scheduled to start before the operation for the component C is finished.

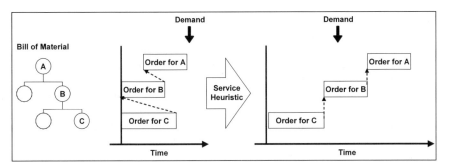

Fig. 4.17 Service heuristics for restoring the material flow

Adjusting the operations to the material flow provides a better starting point for subsequent scheduling heuristics or sequence optimization. For non-bottleneck resources the service heuristics might even substitute other scheduling steps.

Scheduling heuristics are used to schedule the operations into a feasible sequence. The operations are scheduled one after the other, and – if finite planning is used – the order applied to schedule the operations is not necessarily the sequence that results from the scheduling heuristic. The example in Fig. 4.18 shows an operation 4711, which is the first one to be scheduled but is given a later date than operation 4712 because it does not fit into the first free slot.

The order for scheduling the operations is defined by sorting according to one or more criteria, such as the current end date or the set-up group. Standard scheduling heuristics are the rescheduling of selected operations, the rescheduling of operations with backlog, and the scheduling of de-allocated operations. It is also possible to create the order for rescheduling of selected operations interactively.

If the production structure becomes more complex, creating feasible plans might no longer be possible with the simple scheduling heuristics. One approach to problems of this kind is to schedule the whole operation network of one or more orders by multi-level heuristics instead of the individual operations (Dickersbach 2006). An example for a standard multi-level heuristic is rescheduling backlog operations without changing the operation sequences on the selected resources (*cf. desirable feature 2.3.5.3.e*).

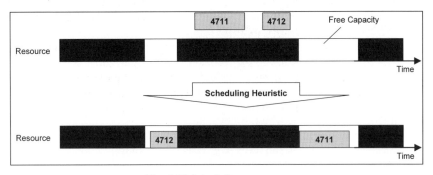

Fig. 4.18 Scheduling sequence

(D) Sequence Optimization

Sequencing problems belong to the class of optimization problems that typically cannot be solved by linear or mixed-integer programming. Therefore, the so-called PP/DS Optimizer applies genetic algorithms that imitate the evolutionary process to find the best sequence of in-house production orders. The objective function for the PP/DS Optimizer considers the criteria lead time, set-up times and costs, average lateness, and maximum lateness. The relative impact of these criteria is determined by weighting factors. Each criterion might tend to favor a different solution, e.g., an optimal solution regarding the lead time might increase the set-up times, and vice versa. It is possible to prioritize demands according to their delivery priority, orders according to their order priority and status (if already begun or confirmed), and alternative resources according to their priority (*cf. desirable feature 2.3.3.2.e*). However, a trade-off between production or set-up costs and storage costs cannot be modeled with the standard settings. It is possible to apply other algorithms for solving special problems, e.g., for trim sheet optimization in the mill industries.

The so-called PP/DS Optimizer is a valuable tool to reduce sequence dependent set-up times, to reduce lead times, or – in the case of inevitable lateness – to schedule orders according to the priority of the demands. And the Optimizer might also have an important role in creating feasible plans at all. On the other hand, the downside of optimization is that its results are not always easy to understand. It is also possible to combine the so-called PP/DS Optimizer and scheduling heuristics, e.g., service heuristics for the rearranging the material flow (see Section 4.2.4.C), so-called PP/DS optimization for the most critical resources, and scheduling heuristics for the other resources.

(E) Modeling of the Physical Production

The properties of the physical production process and their modeling drive the complexity of PP/DS to a large extent. Examples are:

- Time constraints between production steps (within or between orders, e.g., a maximum duration between activities in steel mills because metal has to be processed before solidification).
- Overlapping production (e.g., continuous production of a commodity with a huge lot size and an overlapping packaging).
- Alternative or additional resources (e.g., for labor).
- Resource compatibility for alternative production sequences (e.g., etching and coating can be performed on four alternative resources each, but if etching is done on resource 1, coating has to be performed on resource 2).
- Furnace processes which have two capacity dimensions (e.g., for annealing the furnace offers time capacity for the duration of the annealing process and volume capacity which is consumed by the number and volume of the workpieces).
- Production resources and tools (e.g., dies for molding).
- Batch pureness for production [e.g., for compliance with Good Manufacturing Practices (GMP) in pharmaceutical industries].

- Consideration of shelf life (*cf. desirable feature 2.3.6.3.a*) for food and pharmaceutical products (in SAP APO™ limited to a location product and not integrated with ATP).
- Variant configuration and object dependencies.

(F) Industry-Specific Functions

Since the physical production processes differ from industry to industry, multiple industry specific functions exist in PP/DS. Characteristics of the process industry (e.g., chemicals, life science and some consumer products) are the production of usually large quantities of bulk, e.g., by mixing and heating in a reactor. The content of one vessel is a batch, and sometimes preservation of batch pureness is required in subsequent production. Other functionalities for the process industry are campaign planning and push production in order to process bulk further even without current demand for technical reasons. Another common requirement in this area is scheduling with container resources – either as tank storages or as reactors (*cf. desirable feature 2.3.6.3.c*). However, currently only a very limited functionality is offered for infinite scheduling in a tank storage scenario, with strict allocation of a product to a container resource and with procurement of additional master data.

The planning processes in the mill industries (realizing metal, paper and wood production) are typically characterized by a divergent material flow starting from a coil (metal) or a reel (paper), which is described by multiple characteristics such as strength or grade. The coils or reels are produced to stock, and the characteristics restrict their usage for the finished product. Therefore, propagation of the characteristics from the customer order for selecting the appropriate batch is one of the typical tasks in mill industries. Another task is block planning to group the production orders for coils or reels by characteristics values. Sometimes only a part coil or reel is produced but not a whole one. In this case the production step has more than one output, and the characteristics of the output are propagated from the input using the multiple-output planning functionality.

The cutting of a sheet is a common optimization problem in the mill industry. Although SAP does not offer a procedure for trim sheet optimization there is an interface to third-party tools. Trim sheet orders with special master data are used to model the multiple output after cutting of the trim sheet. For such products as tubes or cables the length is a key parameter for ATP and production planning. Long Products Planning covers the requirements for checking the existing batches for the required lengths, creating cutting orders if necessary, and combining cutting operations.

An important topic for industrial machinery and components is order promising and production planning for highly configured make-to-order or engineer-to-order products with multiple BOM levels and many operations. Therefore, the order-BOM and project networks are transferred from SAP ERP™. Closely related are the use of buffers for infinite planning, heuristics to change the buffers, and a procedure for the display of the critical path.

In the automotive industry the order volume is usually quite high and repetitive manufacturing is used. A specific feature is that the final product – the car – is usually highly configured and contains very many components. The production is usually synchronized, which simplifies the scheduling problem because it levels the differences in the durations of the operations. Production planning and detailed scheduling are performed in one optimization step with Model-Mix Planning (MMP). For MMP the constraints are defined explicitly as entities and are applied at product or characteristic level. For performance reasons the determination of the component requirements is performed in the Rapid Planning Matrix within MMP. Another feature of this solution is that the order confirmation is performed in SAP APO™ for configurable products. MMP cannot be used for non-configurable products. For repetitive manufacturing of non-configurable products there exist production planning heuristics that already consider the capacity requirements of the planned orders. However, the explosion of the BOM on the top level is only done in a second step and the planning of lower BOM-levels, in a third.

(G) Purchasing and Subcontracting
Production planning in SAP APO™ (both in PP/DS and SNP) creates purchase requisitions and selects a supplier according to the ability to meet the requested date, the priority, and the cost. To split the purchase requisitions among different suppliers, quota arrangements are used to overrule priorities and costs. The purchase requisitions are usually converted into purchase order in SAP ERP™. Optionally the conversion can be triggered from SAP APO™, but in this case it is not possible to combine purchase requisitions for the same supplier to give a single purchase order. Scheduling agreements and contracts are also considered in SAP APO™.

SAP APO™ assumes unlimited availability in purchasing, and the only constraint in planning is the procurement lead time (with the exception of collaborative procurement using scheduling agreements and the SNP Heuristic; see Section 4.2.2). Only within the procurement lead time is it possible to regard confirmations as constraints. In cases where the component availability from the supplier is a limiting factor, workarounds have to be applied. One example is a limitation by the volume of a contract, and the restriction might apply to a single product or to a set of products. This restriction is modeled in SAP APO™ by workarounds using transport resources or modeling the production situation of the supplier. The latter procedure is a mere workaround; it has nothing to do with the real production at the supplier, and does not involve collaboration with it.

In subcontracting some production steps are outsourced (*cf. desirable feature 2.3.5.1.b*). In this case the procedure differs from normal procurement that the subcontractor receives the required components from the customer, processes them, and sends the resulting product back. To model this, a purchase requisition or purchase order for subcontracting causes a demand for the components. Examples for subcontracting are products with irregular demand (e.g., displays or kits for promotions in the consumer products industry), production steps that require costly equipment or specialized knowledge (e.g., hardening or electroplating), and production steps with high manual efforts, which can be performed more cheaply in countries with a lower wage level.

In some cases the parts to be provided for the subcontractor are not produced by the customer but procured externally. In this case it might be advantageous to send the component immediately from the supplier to the subcontractor. Another variant is multi-level subcontracting, when the component used by the subcontractor is produced by another subcontractor.

4.2.5 Global Available-to-Promise

The main purpose of the Available-to-Promise (ATP) check is to provide a feasible due date for a customer request. The sales order is created in SAP ERP™, and only the transportation and shipment scheduling and the ATP check are performed in SAP APO™. The result of ATP is the confirmation of the request, i.e., one or more schedule lines where each line contains a confirmed date and quantity. The availability check is a basic functionality mainly for sales, but also for distribution and production. ATP is carried out for the objects sales order, scheduling agreement, delivery, stock transfer order and production order – though not all options exist for every order. Nevertheless, the focus of ATP is sales oriented and operative (Fig. 4.19).

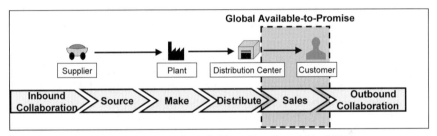

Fig. 4.19 Scope of Available-to-Promise within SCM

The basis for the confirmation depends on the business-specific restrictions and requirements – a confirmation based on available stock only is too conservative in most cases, and therefore planned receipts are often also included. In the case of make-to-order there is by definition no available stock or planned receipt (which has to be created for the request). ATP functionality is also available in SAP ERP™; ATP in SAP APO™, however, does offer the following advantages:

- Global rules-based ATP with substitution of locations and products,
- Global allocations planning through integrated functionality with demand planning,
- Check against a feasible planning result (if planning is performed in SAP APO™),
- Trigger for production by using CTP if there is no availability,
- Check of component availability using multi-level ATP,
- Enhanced check options in backorder processing, and
- High performance using the time series liveCache.

The ATP functionality in SAP APO™ provides three basis methods:

- Product availability check: check of free stocks or receipt elements (level product and location).
- Forecast check (level product and location). The assumption in this case is that everything that has been planned is also procured.
- Allocation check: check of allocation quantities (any level) in order to guarantee customers a share of the products in case of shortages.

The product check is the most commonly used ATP basis method (*cf. desirable feature 2.3.3.2.a*), and the calculation of the availability is performed anew for each check using daily time buckets – there is no fixed correspondence between receipt and requirement elements. Alternatively it is possible to carry out the ATP check using the pegging network – e.g., if a link between a sales order item and a receipt should be kept via fixed pegging during planning (see Section 4.2.4.A).

(A) Allocations
The functionality of the SAP object "allocations" differs slightly from the usage of allocations in everyday speech. Allocations help to ensure that customers or customer groups are each provided with their share according to the sales and marketing policies in cases where the demand or the expected demand exceeds the supply. Generally, allocations prevent the application of the "first-come, first-served" rule. Allocations are not reservations of stock or planned receipts, but are used to limit the confirmation of the customer demand in the ATP check. The effect of reserving quantities for customer groups is achieved by preventing other customers from claiming more than their share. If allocations are used for a product it is therefore necessary to allocate all supplies of the product (and not only a portion designated for certain customers). Another motivation for applying allocations is the opportunity for checking against limited production capacities, assuming that the quantity needed can be produced later, if necessary. It is possible to perform allocation checks for a sales order item at several levels, e.g., customer group and sales organization. The results can be combined either as an intersection (minimum) or as a union (sum).

One of the advantages of the allocation solution in SAP APO™ is that the allocation quantities are planned in Demand Planning (see Section 4.2.1). It is possible to connect allocation planning in DP online to ATP, which is an advantage if the allocation quantities are changed frequently and an immediate impact is desired.

(B) Rules-Based ATP
An ordinary ATP check is restricted to the requested product at the delivery plant. Using rules-based ATP it is possible to substitute both the location (*cf. desirable feature 2.3.3.2.b*) and the product (*cf. desirable feature 2.3.3.2.c*). In supply chains where mostly uniform products are kept at multiple locations or where different product numbers are sold to customers as identical products, rules-based ATP is a

valuable tool to increase the delivery performance and to reduce the safety stock levels. The limitations for the use of rules-based ATP are explained elsewhere by Dickersbach (2006).

In the case of a location substitution there are two options (Fig. 4.20). As an example, a sales order item is assigned to the delivery plant XX01, where the request is checked first but rules-based ATP determines a confirmation from location XX02. The two options in this case are:

1. The delivery plant for the sales order item is changed to XX02. The customer is then supplied from location XX02 instead of XX01.
2. The delivery plant for the sales order remains XX01 and a stock transfer requisition is created from location XX02 to XX01. The customer is supplied from location XX01.

Location substitution is performed for each sales order item. Using the rule type for multi-item single delivery location (MISL), all items of the same delivery group are sourced from the same location. The standard logic is that as soon as an item is not fully confirmed the next location is checked for all items. Another feature in this context is the use of a consolidation location in order to consolidate partial confirmations from multiple-location substitutions before they are sent to the customer.

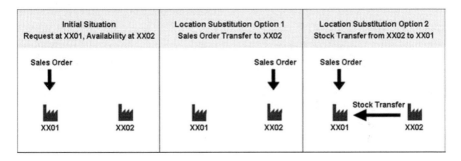

Fig. 4.20 Location substitution with and without stock transfer requisition

(C) Capable-to-Promise
Capable-to-Promise (CTP) triggers PP/DS to create planned receipts during the ATP check in order to ensure product availability (*cf. desirable feature 2.3.3.2.a*). At planned order creation the availability of the capacity and of the key components is checked. Therefore, CTP is especially helpful if the availability date has to be confirmed to the customer immediately. This functionality is frequently used especially in the make-to-order environment of mechanical engineering companies, but can also be used for make-to-stock. There are several steps during a CTP check that involve both ATP and PP/DS functionalities (Fig. 4.21).

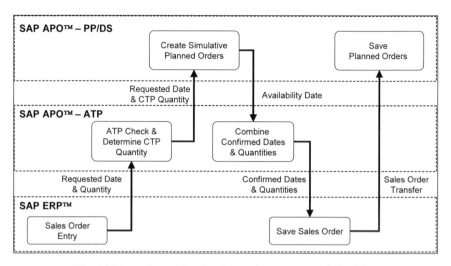

Fig. 4.21 Procedure of the Capable-to-Promise check (simplified)

First in ATP the requested date and the requested CTP quantity are calculated. PP/DS is called with this information and creates a so-called simulative planned order according to the production restrictions. PP/DS is always able to create planned orders for the requested quantity, even if it might be very late (if production is triggered from rules-based ATP it is possible to restrict the lateness using a parameter of the rule). Now PP/DS returns with the availability date for the CTP quantity and ATP combines the confirmations from the ATP check with the PP/DS result. After saving, the sales order is transferred to SAP APO™ and the planned order is saved. The procedure is explained in more detail by Dickersbach (2006).

The capacity check for the planned order is performed using either the detailed, time-continuous capacity or an aggregated bucket capacity. The time-continuous capacity check implies time-continuous finite scheduling of the planned orders, which is usually sub-optimal in terms of sequence-dependent set-up and scattered loading and therefore leads to low resource utilization. It is recommended to use the aggregated bucket capacity to avoid this problem. In this case the planned orders resulting from the CTP check are scheduled "bucket finitely," but within the bucket infinitely (without considering capacity constraints). Nevertheless the use of CTP implies the creation and scheduling of one or more orders per check, and the complexity increases with the number of operations, the number of finite resources and the number of BOM levels. Therefore there should be only one finite resource for the CTP check per BOM level. Examples of other limitations are scheduling agreements, the combination with lot sizes, and the use of the so-called requirement strategies "planning without final assembly" and "planning product." In any case a subsequent scheduling step is required after CTP.

(D) Multi-Level ATP
Multi-level ATP confirms a request if the components needed for the product are available (taking the lead time to produce the finished product into account). Multi-level ATP is mostly used in cases where the finished products are assembled to order and the production capacity is not a bottleneck. In this case the availability of the finished product depends on the availability of the components. The primary focus of multi-level ATP is make-to-order production, but make-to-stock is also supported. The discrepancy between the bucket-oriented ATP check and the time-continuous production might become an obstacle to the use of multi-level ATP if many BOM levels are checked, e.g., in the engineering and industrial machinery industries. This is because for ATP the production duration is a multiple of a day – this way the differences to the real production duration might add up, with the consequence of confirming a late date. The production capacity is not considered during multi-level ATP, and a subsequent scheduling in PP/DS is required.

(E) Characteristic Based ATP
In the mill industry products are often described by many characteristics that determine their disposal. But also in other industries the result of the production might differ depending on uncontrollable factors, especially in chemical production processes. In these cases it depends on the characteristic values whether a request can be confirmed or not: for example, if the customer requires steel sheets of a thickness between 0.03 and 0.031 inch, it is not acceptable to confirm this demand by citing the availability of steel sheets with a thickness of 0.029 inch.

Unless the business function group for discrete industries and mill products (DIMP) is used in SAP ERP™, this scenario allows sales exclusively from stock, i.e. no planned receipts are considered. This is a hard restriction, because there is no interface for transferring planned receipts data with characteristic values in the make-to-stock segment to SAP ERP™.

The sales order receives its requirements for the characteristic via the batch selection. Unlike the common way of using batch selection, here only the characteristics of the batches are evaluated (and transferred to SAP APO™) and not the batch numbers. Within the ATP check only those elements are used for confirmations that have a configuration that matches the requirements of the sales order. Usually rules-based ATP is used to allow a substitution of the characteristic values (e.g., if multiple values or an interval are maintained as batch selection criteria). This ensures that the sales order has only a single value (or none if it is not confirmed).

(F) Transportation and Shipment Scheduling
The transportation and shipment scheduling is an integral part of the availability check in SAP APO™ and calculates the time difference between the requested delivery date at the customer site and the required material availability date at the factory. Scheduling is performed using a set of fixed values for transportation planning, picking & packing, loading, transportation, and unloading, which are determined in a flexible way, or by using the functionality of the Dynamic Route

Determination (see Section 4.2.6.B). If SAP APO™ is used for the ATP check, the transportation and shipment scheduling functionality of SAP APO™ has to be used instead of the analogous functionality in SAP ERP™.

(G) Backorder Processing
The current availability situation might differ significantly from that planned at the time of the sales order. This means that the current confirmation of dates would be revoked. One main motivation for resolving such a situation is the need to enhance or restore the visibility and the clarity of the planning situation. If it is clear that the confirmed dates are not met, unrealistic requirement dates cause alerts for shortage or lateness and misleading goals for production planning. Nevertheless, typically the first goal aspired to is coverage of the confirmed demands. Therefore backorder processing should not be performed until production planning is finished. A second point is that in many cases, especially when the lead time between placement of an order and the requirement date is long, the customer should be informed about the delay. Whether information to the customer, e.g. by sending a new confirmation, is required, depends on the sales order process. Sending confirmations to customers is sited entirely within sales in SAP ERP™; it can be triggered by a change in the confirmation. A new confirmation, however, does not necessarily cause a notification to the customer.

The basic idea of backorder processing (BOP) is to carry out a new ATP check for a set of order items in a certain sequence. This sequence defines how supplies are allocated to the sales order items in the case of shortage or lateness. Additional parameters influence release of the confirmations during the check, whether the request or the confirmation is checked, and whether the rules are re-evaluated (*cf. desirable features 2.3.3.1.s, 2.3.3.2.d, and 2.3.5.3.e*). It is not possible to trigger production via CTP in backorder processing. It is, however, possible to include products that are planned with CTP in the BOP and check these without creating new planned orders. The complete backorder process contains the steps backorder processing in SAP APO™, transfer confirmation to SAP ERP™, and update sales order in SAP APO™.

The planner can overrule the result of the ATP check or the BOP run by changing the confirmations of sales order items interactively. It is possible to confirm a higher quantity than available, though in this case the system provides a warning message. The only restriction is that it is impossible to exceed the requested quantity.

An alternative approach used to deal with backorders is realized with the order due list (ODL). The planner includes only a few sales orders of high importance into this list at the beginning of the day and the change in the availability during the day – e.g., receipts from production or distribution – are monitored and assigned to the sales order items. The quantity assignment is done interactively for the selected order items.

4.2.6   Transportation Planning and Vehicle Scheduling

The purpose of transportation planning is the combination of freight units (deliveries, sales orders, or stock transfer orders) into shipments. A shipment specifies which freight units are combined in which means of transport (e.g., big truck, small truck), and the sequence in which they are delivered. Obviously there is an optimization potential by combining the freight units and selecting the route for the shipments in order to minimize the effort – i.e., the number of shipments and the length of the shipments – while taking account of the due dates, the delivery windows of the customers (for loading and unloading) (*cf. desirable feature 2.3.3.2.g*), the capacity restriction of the vehicles, the vehicle availability (*cf. desirable feature 2.3.3.2.f*), and incompatibilities. Shipments are usually created daily or several times per day. In most cases TP/VS is used for the outbound planning of shipments to the customers but it is also possible to plan inbound shipments from the suppliers or stock transfers within the internal supply chain. Therefore, TP/VS is a tool for operative planning mainly in the sales area (Fig. 4.22). Within SAP APO™, transportation planning for distribution is typically performed by TLB (see Section 4.2.2.J).

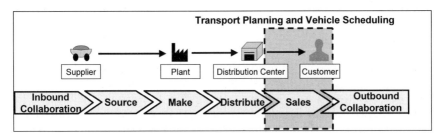

Fig. 4.22 Scope of Transportation Planning within SCM

The result of transportation planning is the creation of a shipment in SAP APO™. This document has a link to the related deliveries or sales orders but does not replace them. Accordingly a shipment is not relevant for requirements planning.

TP/VS is designed for the transportation planning of a producing or a trading company but less appropriate for third-party logistics providers (3PL). Though it is possible to plan based on freight units that are maintained in SAP ERP™ with a general number (e.g., order for material "pallet"), in most cases 3PL have requirements in terms of driver planning and multi-partner tendering which exceed the functionalities of TP/VS.

Subsequent process steps are selection of the carrier and release of the shipment. The shipment is not transferred to SAP ERP™ until the release in SAP APO™. With the transfer a shipment is created in SAP ERP™. Fig. 4.23 provides an overview of the order statuses.

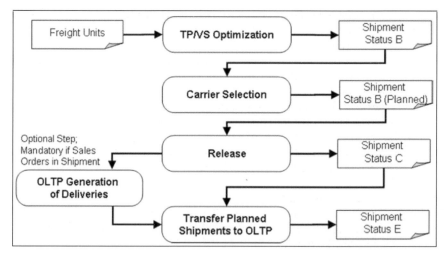

Fig. 4.23 Order statuses of shipments in Transportation Planning

Deliveries are created in SAP ERP™ before TP/VS is run, or alternatively triggered from TP/VS. In this case TP/VS plans on a sales order basis. The ATP check for deliveries usually has a more restricted scope than the sales order and therefore the delivery might not get the full confirmation which implies that the shipments have to be corrected in SAP APO™.

(A) TP/VS Optimization
The task of the so-called TP/VS Optimizer is to create a plan (i.e., shipments) which considers the constraints and results in low penalty costs. Hard constraints for the optimizer are incompatibilities (cf. Section 4.2.2.J), opening hours (modeled by the handling resource), and the finite capacities (as defined in the resource master data). Pick-up and delivery windows may result in hard constraints (*cf. desirable feature 2.3.6.2.b*). Soft constraints are defined in the optimizer profile (e.g., allowances for earliness and lateness).

The procedure for the so-called TP/VS optimization is to begin with some start solutions and successively load further orders. After a while a few of these solutions are taken into account for further processing, and the final result is the best solution of the whole cycle. Evolutionary algorithms are used – the TP/VS Optimizer applies local search and evolutionary search techniques. The complexity of the problem is driven by the number of means of transport, hubs, handling resources, and orders. The optimizer considers costs of delayed and premature delivery and pick per day, costs per vehicle (one time cost), costs for distance, quantity, and duration, and costs per stop off.

With the concept of (in)compatibilities (possible between any kinds of fields and not only between transport groups), constraints on combining freight units can be considered in a flexible way (*cf. desirable feature 2.3.6.3.f*).

Calculation of the transport durations depends on the accuracy of the geo-coding of locations and of the distance between them. The determination of the geographical settings (longitude and latitude) of the locations and the distance of the transportation lanes based on the address and the actual distance between addresses using a route planning requires geo-coding software, which is not included in the standard license. Alternatively it is possible to determine the geographical settings with less accuracy (included in the standard license).

The so-called optimization is performed in the background or in the TP/VS planning board. There it is also possible to display the capacity load of the vehicles (*cf. desirable feature 2.3.6.3.d*), to assign freight units interactively to shipments, to change shipments, or to apply customer specific heuristics.

(B) Dynamic Route Determination
Dynamic Route Determination uses pre-defined routes in order to provide a more precise transportation and shipment scheduling during the ATP check of sales orders (or quotations). With this method, schedules and resource capacities are already considered during the ATP check, and a shipment is created. This shipment is likely to remain unchanged only in the case of full truck loads, or alternatively TP/VS might change it during optimization. It is possible to display different transportation alternatives during the ATP check, sort them by costs or tardiness, and select one interactively. The routing guide is also used for interactive scheduling in the TP/VS planning board.

(C) Carrier Selection
The carrier selection is performed after the planning of the shipments is finished and before the shipments are transferred to SAP ERP™. Criteria for carrier selection are priority, costs, or allocation. If a carrier is not available or declines the offer, the next one is chosen. If a route contains several stages, only those carriers are considered that are assigned to all lanes.

It is possible to communicate the result of the carrier selection in a collaborative scenario. In this case the requests for the shipments are sent to the transport service provider via the Internet or via ALE. The requests for the shipments can be accepted or rejected by the transport service provider; a rejection creates an alert.

4.2.7   Maintenance and Service Planning

The maintenance and service planning (MSP) solution is designed for airline companies and differs from typical supply chain planning in several ways. However, since MSP is a part of SAP APO™, it is explained for the sake of completeness. Its purpose is to plan the capacity requirements for maintenance services. Only the capacities for the bay or hangar and for the labor are planned; the material requirements for consumables and rotables (e.g., change of an engine) are not planned in MSP but in SAP ERP™.

The services differ in scopes and effort and are subject to many legal requirements. A maintenance check with a certain scope (e.g., A-check, C-check, overhaul) is due after a certain number of flight hours, cycles (i.e. start and landing), or days. The due date for the maintenance check is calculated on the basis of the current counter of the aircraft (which is updated from SAP ERP™) and the average flight rate. The due date is also the same as the compliance date, because the aircraft is not allowed to fly any more if the check has not been performed. The challenge for maintenance and service planning is to plan the checks before the compliance date but not too early, in order to keep the average utilization of the aircraft high. Another challenge is combining different checks for the aircraft in order to have only one downtime for the aircraft. The limiting factor is the bay or hangar availability for the maintenance checks, but the labor availability of the different qualifications also has to be planned.

For the different specifications of the checks (e.g. A, 2A, or 4A) either different maintenance plans or the cycle set sequence is used. In MSP the check type is used as a criterion to package checks (e.g., check type "A" for the checks A, 2A, and 4A). The task lists are not transferred from SAP ERP™ but have to be maintained in SAP APO™.

The MSP planning run creates the demands for the maintenance, the check orders (representing the tasks for the workers), and the slots (representing the downtime of the aircraft at the bay or hangar), based on the cycles. The demand for the maintenance object is synchronized with the calls in SAP ERP™ that have the same semantic. The check orders in SAP APO™ correspond to the notifications in SAP ERP™. However, the notifications are created in SAP ERP™ by the release call, and only the schedule is updated from SAP APO™. The slots on the other hand correspond to the revisions. Assignment of the check orders and the slots is the work package in SAP APO™. For the combination of different checks it is possible to assign several check orders to one slot. Work packages are created interactively by the planner and transferred to SAP ERP™ as the assignments of notification to the revision. Fig. 4.24 shows the order flow for Maintenance and Service Planning.

Because the actual use can differ from the estimated use (e.g., due to rerouting and technical problems), the calculation of the checks is repeated on a regular basis. Each check has a compliance date by which the check has to be done. The latest start date for the check includes the check duration and a safety buffer – there might be delays in the schedule or unexpected repairs might be required. The earliest start is an economical boundary – the earlier the check is performed, the lower the average utilization of the aircraft. Somewhere in between is the preferred start date, where the system tries to schedule the check.

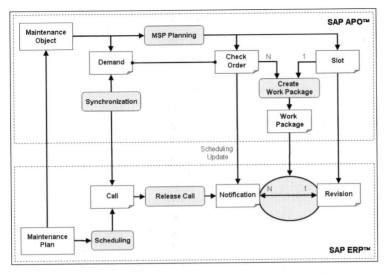

Fig. 4.24 Order flow in Maintenance and Service Planning

The dates are calculated based on the cycle duration (which is calculated as the ratio of the cycle – defined in the maintenance plan – and the annual estimate times 365 days) and the tolerance profile assigned to the cycle in SAP APO™ (Fig. 4.25).

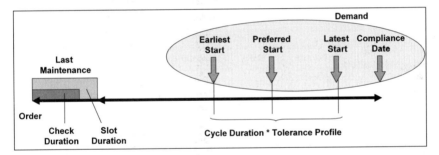

Fig. 4.25 Scheduling of the maintenance checks

The slot duration means the length of time the aircraft is in the bay or hangar and is derived from the slot task list. This list contains also the feasible bays or hangars as alternative resources. The slots are scheduled finitely within the tolerance profile. If no bay or hangar is available within the limits, the slot is not scheduled by the system and the planner has to schedule the slot interactively. The check duration reflects the labor requirements for the check and is derived from the check task list. The labor requirements are scheduled without considering capacity limits. In all cases it is possible to change the schedule interactively in the MSP planning board.

4.3   Collaboration supported by SAP ICH™

The Inventory Collaboration Hub (SAP ICH™) is a dedicated hub for collaboration with an adjacent supply chain partner, i.e. the supplier or the customer. SAP ICH™ contains two separate components for collaboration with suppliers (inbound) and for collaboration with customers (outbound); the latter is called Responsive Replenishment (Fig. 4.26).

SAP ICH™ has been renamed to Supply Network Collaboration (SAP SNC™) recently and is also available as a separate product (i.e., without other SAP SCM™ products).

The components for Supplier Collaboration and for Responsive Replenishment are completely disjunctive. The purpose of the supplier collaboration is to offer small and medium-sized suppliers a low-price collaboration platform via the Internet. Currently EDI is only used for communication with 20% of the suppliers (Leitz 2005). Responsive replenishment has the industry-specific focus on collaboration between a consumer product company and its customers, the retailers. SAP ICH™ contains a web-based user interface; therefore it is possible to access SAP ICH™ via the Internet without using a SAP GUI. Throughout all application systems and processes, alerts notify the planner of relevant exceptions.

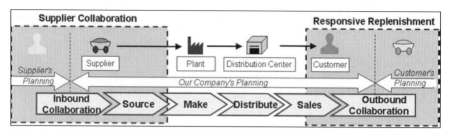

Fig. 4.26 Scope of Supplier Collaboration and Responsive Replenishment within SCM

4.3.1   Supplier Collaboration

Supplier collaboration based on SAP ICH™ enables different collaborative processes via the Internet. There are two types of collaboration – one is transferring the management of the inventory to the supplier and the other the technical transmission of orders and their confirmation via the Internet. SAP ICH™ is used for purchase orders, scheduling agreements and releases, Kanban requests, and subcontracting purchase orders. Depending on the process and the order type, different monitors are used for planning. The supplier collaboration processes with SAP ICH™ are intended for the case where the customer uses SAP ERP™ and the supplier system has no interface to the customer system. In this case the customer's data – purchase orders, releases, gross requirement and/or stock (depending on the process) – is sent from SAP ERP™ to SAP ICH™. The integration is realized via SAP XI™ (Exchange Infrastructure), which is part of SAP Netweaver™ and is installed as a separate system. In SAP ICH™ both customer and supplier can access the data via the Internet. The processes for technical transmission of orders (purchase orders, scheduling releases, Kanban requests, or subcontracting purchase orders) are not different from those in SAP ERP™ (with the exception that the supplier can confirm the scheduling releases in SAP ICH™). However, compared with sending of purchase orders or releases by fax, SAP ICH™ offers the advantage of avoiding paper and keeping track of the order history. The advantage over EDI is that only access to the Internet and no costly infrastructure is required.

(A) Purchase Order Collaboration
In the purchase order collaboration process, SAP ICH™ is used as a technical platform to transmit purchase orders to the supplier. The customer creates a purchase order in SAP ERP™. This order is transferred via SAP XI™ to SAP ICH™ and triggers a notification alert for the supplier. The supplier confirms the purchase order depending on availability. The customer receives an alert if the purchase order is not confirmed after a predefined time. When the product is shipped, the supplier creates an Advanced Shipping Notification (ASN) in SAP ICH™. Both the confirmation and the ASN update the purchase order in SAP ERP™. Fig. 4.27 shows the purchase order process. The process step "Close ASN" which is underlaid with gray is performed in the background.

Since the data in SAP ICH™ contains only logistical information (and no business information), the purchase order in SAP ICH™ does not have the same legal status as a "regular" purchase order.

(B) Release Collaboration
The release collaboration is analogous to the purchase order collaboration but uses scheduling agreements and releases instead. The customer creates a scheduling agreement and releases in SAP ERP™ and sends the delivery schedule to SAP ICH™, where the supplier is informed via alerts about new releases. With scheduling agreements it is also possible to plan and create the releases in SAP APO™ and use SAP ICH™ for collaboration. An advantage of this process compared with release creation in SAP ERP™ is that the supplier can confirm or reject the releases (*cf. desirable feature 2.3.5.3.a*). The confirmation of releases uses the CMDS functionality of SAP APO™ (see Section 4.2.3.D).

(C) Dynamic Replenishment
The Dynamic Replenishment functionality can be used both for the purchase order collaboration and for release collaboration. It enhances these processes by automatic upload of the product availability at the supplier's site and its comparison with customer requirements. Both partners are informed by alerts if there are mismatches (*cf. desirable feature 2.3.5.3.a*). Integration of the availability at the supplier via SAP XI™ requires more effort, because there is no standard business content in SAP XI™ for this data. This process makes sense only if the products are customer specific; otherwise the available products could be used to satisfy the demands of other customers.

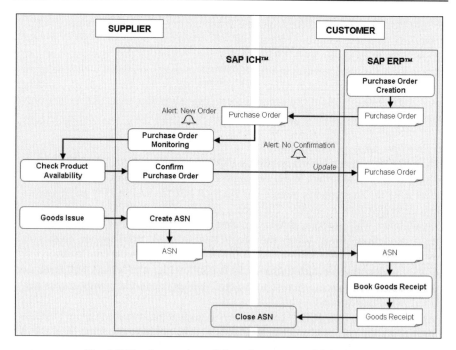

Fig. 4.27 Purchase order collaboration process with SAP ICH™

(D) Supplier Managed Inventory
In SAP terminology a distinction is made between Vendor Managed Inventory
(VMI) and Supplier Managed Inventory (SMI). In the case of SMI the system is
managed by the customer. In both cases the planning is done by the supplier. The
customer provides the supplier its stock data and its gross requirements to the
supplier, and the supplier replenishes the customer's warehouse within minimum
and maximum stock levels. The advantage for the supplier is the higher flexibility
to synchronize the delivery with its own product availability and capacity utilization, and the advantage for the customer is less planning effort. There are two
process alternatives for SMI, either using purchase orders or scheduling agreements:

- The supplier creates purchase orders in SAP ICH™. The purchase orders are
  transferred from SAP ICH™ to the SAP ERP™ of the customer.
- The customer creates a scheduling agreement and a schedule release (with
  a target quantity which would match the yearly requirements).

In both cases the supplier creates ASN, and goods receipt is booked in the
customer's SAP ERP™ (*cf. desirable feature 2.3.3.1.f*). Fig. 4.28 shows the process
flow for SMI with scheduling agreements.

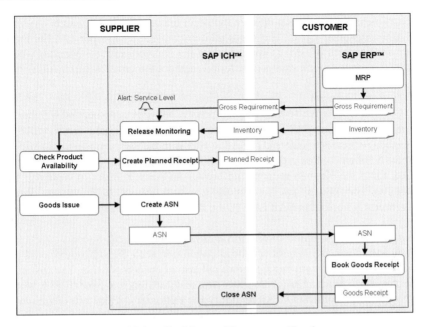

Fig. 4.28 Supplier Managed Inventory with releases

When using purchase orders for SMI, the supplier must create a purchase order in addition (analogous to the purchase order collaboration process) before creating an ASN. Table 4.2 lists the advantages and disadvantages of the alternatives (Leitz 2005).

Table 4.2 Alternatives for SMI

	Advantage	Disadvantage
SMI with purchase orders	Purchase orders visible in SAP ERP™ No scheduling agreements required	More effort and responsibility for the supplier (problems at goods receipt if purchase order is not created on time)
SMI with releases	Fewer process steps, less maintenance for supplier	No receipt elements in SAP ERP™ (with implications for ATP)

SMI requires trust between customer and supplier and is only usual in the case of a close single sourcing relationship. Technically it is also possible to apply SMI for multi-sourcing relationships by splitting the demand according to quota arrangements and splitting the inventory by supplier. However, splitting the inventory requires batch management, which is very unusual for SMI.

Sometimes (e.g., in the automotive industry) the supplier has a warehouse very close to the customer's plant and sends multiple replenishments per day. For these cases a special form of SMI is applied using the Delivery Control Monitor where only the customer's stock data is transferred (and not the gross requirements).

(E) Web-Based Kanban

SAP ICH™ enables a Web-based Kanban-process between customer and supplier. If a Kanban is set to "empty" and a new Kanban request is created in SAP ERP™, this Kanban request is transferred to SAP ICH™ with the corresponding information about the Kanban control cycle and the quantity of the Kanban bin. The supplier is notified about the Kanban request and creates an ASN in order to replenish the Kanban bin. The status of the Kanban changes from "empty" to "in transit." When goods receipt is booked in SAP ERP™, the status of the Kanban changes to "full."

(F) Supply Network Collaboration for Subcontracting

For subcontracting, the benefit for the collaboration with the subcontractor is the visibility not only of the procured product but also of the components. The customer creates the subcontracting purchase order item and transfers it to SAP ICH™, where the subcontractor confirms it. Alerts notify the customer if there is a deviation in the confirmation or no confirmation at all. The confirmation updates the subcontracting purchase order in SAP ERP™, and the customer issues the components to the subcontractor. In SAP ERP™ the components are booked into the stock at the subcontractor's site and transferred to SAP ICH™. On starting with the production, the subcontractor confirms the consumption of the components in the subcontracting purchase order and may change quantities or components. Optionally the customer can approve these changes before updating the purchase order in SAP ERP™. After the subcontractor has issued the product it creates an ASN in SAP ICH™, which is transferred to SAP ERP™ and used for goods receipt at the customer's premises. With the goods receipt of the product the confirmed components of the subcontracting purchase order are backflushed from the stock at the subcontractor's site. Fig. 4.29 shows an overview of the process.

If the customer does not provide all components to the subcontractor but some components are procured by the subcontractor itself, the availability of these components is also of interest for the customer. It is also possible to display the inventory for these components in SAP ICH™. The data is either maintained in SAP ICH™ interactively or integrated via SAP XI™, as described for dynamic replenishment.

An increasingly popular behavior of the customer is to procure components from a third-party supplier and send them directly to the subcontractor (cf. also Section 4.2.4.G). For this process SAP ICH™ increases the visibility by displaying the ASN for the components. Currently the third-party supplier cannot yet join the collaboration in SAP ICH™.

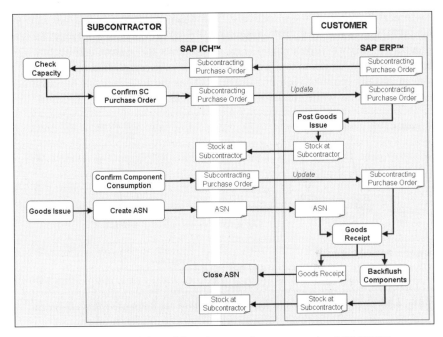

Fig. 4.29 Overview of the subcontracting process using SAP ICH™

Current limitations of the subcontracting collaboration include the unavailability of any BOM information in SAP ICH™ and the impossibility of distinguishing between products and components; in addition, the supplier's inventory of component stock is not updated until after the customer has received the goods, which means that from goods issue of the components to the supplier for production and goods receipt of the finished product by the customer on its site the record of current component stock is not accurate, and it is necessary to use different monitors to display the subcontracting purchase order and the stock situation.

4.3.2 Responsive Replenishment

Responsive Replenishment (RR) within SAP ICH™ has the focus on the consumer product industry for the VMI process between a consumer product company and its retailers (*cf. desirable feature 2.3.3.1.f*). The assumption for this solution is a high-volume business with daily replenishments and unlimited availability at the vendor's site. Based on this assumption, no stock information about the vendor's own DC is available. The objective of RR is planning the replenishment to the customer's (i.e. the retailer's) DCs taking short-term demand changes and transport restrictions (e.g., rounding) into account. The collaboration with the customer is done via EDI and not via the Internet because the collaboration partners are typically large companies. The customer sends the stock and sales data and, optionally, a forecast to SAP ICH™. RR in SAP ICH™ is used exclusively by the vendor in order to calculate the transports to the customer. The transports are sent to SAP ERP™ as sales orders for execution and delivery. The planned stock transfers are sent to SAP APO™ (via the SAP BI™ structures of SAP APO™) as a forecast for the planning of the own supply chain. Fig. 4.30 shows the system architecture for RR. The communication to and from SAP ICH™ is accomplished via SAP XI™. From an IT-safety point of view this architecture has the advantage that the customer has no direct access to the vendor's planning system.

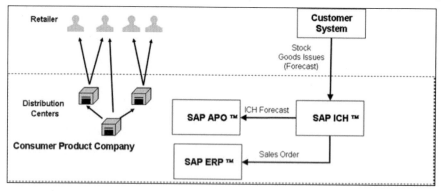

Fig. 4.30 Supply chain structure and rough data flow for Responsive Replenishment

Fig. 4.31 shows the process steps for RR; the two steps underlaid with gray are performed in the background. The stock and sales data is sent to SAP ICH™ and checked for completeness and plausibility. Unless the customer sends its own

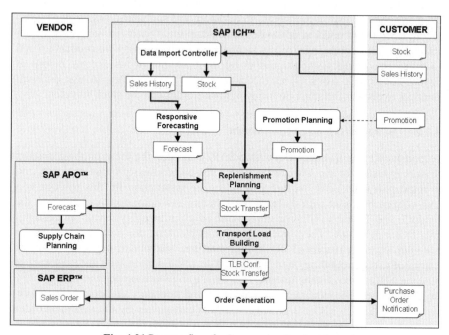

Fig. 4.31 Process flow for Responsive Replenishment

forecast, one is produced by the supplier in the next step. Replenishment creates the unconstrained planned transports to the customer, and transport load building applies the constraints for rounding and truck capacity. Replenishment planning for promotions is done separately. The stock transfers of the planning horizon are sent to SAP APO™ as forecasts and the stock transfers of the next days to SAP ERP™ as sales orders.

(A) Data Import Controller
The purpose of the data import controller is avoiding planning mistakes that could result from wrong data. The main tasks are checking whether the data is complete and within the expected thresholds. The planner is informed by alerts if inconsistencies are detected. An example of an inconsistency is sales for a product which is already withdrawn from the market. In this case the planner corrects the data interactively. The simulation functionality of the data import controller allows using an expected input for planning if that data element from the customer is missing.

(B) Responsive Forecasting
Forecasting is based on the sales history of the customer. The forecast is calculated for weekly buckets with the standard models for constant, trend, seasonal, and seasonal trend behavior. Alternatively it is possible to import the customer's forecast – if available. The weekly forecast is disaggregated to days based on a weighted ratio of the last week's disaggregation factors and the last week's sales.

In the next step the forecast is adjusted for the current week in "short-term forecasting." The projected sales are calculated by scaling the forecast with the ratio of the cumulated sales and the cumulated forecast of the current week. If the deviation between projected sales and forecast exceeds a threshold, the projected sales are used for planning. This way the forecast of the next days is adjusted to a difference in the consumers' demand.

Life cycle planning for the introduction of new products and the phase-out of old products is done based on a scaled reference to other products – like in SAP APO™ Demand Planning (see Section 4.2.1).

For the handling of complex shipping units (CSU) such as displays and promotional packages it is possible to display the dependent demand. The dependent demand is not, however, used for planning, but only for information purposes. The precondition for its use is that production master data for DP exists in order to represent the BOM (see Section 4.2.1.I).

Forecasting is performed at product and location (in this case at customer) levels; it is not possible at an aggregated level (e.g., product group). It is also not possible to use macro functionality or to visualize the results graphically. Forecasting the customer's demand is done based on the goods receipt date at its warehouse and not based on the order date. However, the forecast data that is passed to SAP APO™ for supply chain planning is derived from the transports and is therefore just as lumpy as without RR.

(C) Promotion Planning

Promotions are maintained as quantities with a reference to an expected sales pattern. Promotion data is either maintained interactively by the planner of the vendor or is sent by the customer (*cf. desirable features 2.3.3.1.h and 2.3.3.1.p*). If the promotion is sent by the customer the reference to the sales pattern ID must be included. Since the pattern IDs are vendor specific and arbitrary this requires an increased effort from the customer. RR in SAP ICH™ offers three categories of promotions:

- Static promotions without tracking or updates. Both the total quantity and the pattern are fixed.
- Reactive promotions are updated depending on the initialization periods and the thresholds of the promotion profile. The pattern remains fixed and the total quantity is adjusted.
- Dynamic promotions release the next promotion demand only if a threshold is reached in order to avoid overstocking. In this case the total quantity is fixed but the demand pattern is adjusted.

In addition to the sales pattern, the distribution pattern defines replenishment agreements with the customer, e.g., forward buying to accumulate stock for the promotion. The cannibalization effect of the promotions is considered in promotion planning. It is possible to use a BOM (as complex shipping units) to create the cannibalization group. However, no reports are provided on promotions.

The products for promotions are either separate products with different product numbers (e.g., kits, special packages) or "normal" products. RR in SAP ICH™ requires a distinction between the sales data and stock for normal and promotional demand. This distinction is usually very difficult for sales data, requiring estimations, and it is impossible for stock; it might be done in the Data Import Controller based on estimations.

(D) Replenishment Planning

The purpose of replenishment planning is to calculate the net requirement at the customer location taking order sizes and safety stock into consideration. The safety stock can be defined in three ways: as an absolute value, computed from the replenishment time, or calculated with advanced methods, as in SNP (see Section 4.2.2.A). The result of replenishment planning is planned stock transfers from the vendor to the customer.

Replenishment planning is performed separately for the baseline demand (i.e., the "normal" demand without promotions) and the promotion demand. For each promotion a distinct replenishment planning is performed using separate, promotion-specific stocks. To a certain extent it is possible to use combined stock transfers for baseline and promotion demand, and to use the baseline stock for promotion demand. Nevertheless, it is unusual to separate the stock into baseline demand and promotion demand unless different product numbers exist for them. Therefore this approach might involve some difficulties in implementation.

For replenishment of the promotion demand it is possible to define rules for sourcing – e.g., based on order sizes. The availability at the vendor's locations is not considered, since unlimited availability is assumed and no information exists about stocks and planned receipts of the vendor.

(E) Transport Load Building
The planned stock transfers from replenishment planning are combined into feasible stock transport with due consideration for the constraints of the means of transport. The functionality of the transport load builder (TLB) is almost the same as in SAP APO™ TLB (see Section 4.2.2.J); exceptions are a prioritization of the promotion demand and some specific settings for the promotion demand, such as prevention of downsizing the order quantity of the shipment or selection whether or not goods for baseline and for promotion demand might be mixed. Another feature is the consideration of costs in up- and downsizing (the cost data has to be maintained interactively in SAP ICH™). The constraints are maintained in the TLB profile per transportation lane. Another difference from SAP APO™ is that it is possible to maintain the constraints and rounding parameters per transportation lane and product in a transportation guideline. The results of TLB are TLB-confirmed stock transfers.

(F) Dynamic Sourcing for Promotions
The purpose of dynamic sourcing is to reduce the transportation costs by re-plenishing huge quantities for promotions directly from the plant. In the case of dynamic sourcing the replenishment planning is combined with transport load building. When planning a location product, the transportation lane from a source with the transportation guideline of the highest priority is selected. All products of a transportation guideline are planned, and a check is made on whether all con-straints for TLB are met. If they are not, the products of the transportation guideline with the next priority are planned, and so on. The result of this procedure – TLB-confirmed planned orders – depends therefore on the planning sequence of the products (which cannot be influenced). The dynamic sourcing functionality is only possible for promotion demands.

(G) Order Creation
The TLB-confirmed stock transfers for the current day or the following days are released manually and sent to SAP ERP™. The sales order number is created in SAP ICH™ and assigned to the SAP ERP™ sales order. Only the released TLB-confirmed stock transfer orders are regarded as fixed planning elements. If one sales order contains a demand for the same product resulting from baseline demand and from promotion demand, two different sales order items are created. With the release to SAP ERP™ a purchase order notification is also sent to the customer. TLB-confirmed stock transfers for the medium-term horizon are sent to the SAP BI™ part of SAP APO™ and used as forecasts.

RR has some valuable features, such as daily adjustment of the forecast bucket and direct shipment based on lot sizes. Nevertheless, the solution also has some restrictions, such as the lack of interactive planning, lacking visibility of the own stock and availability situation and therefore no balancing of own shortages, and lack of forecasting at aggregated level. In our experience the separation of stock for normal demand and promotional demand is not realistic and might lead to problems if the same product number is used for baseline demand and promotions. The comparatively low adoption of this solution indicates that solutions exist with a more favorable ratio of benefit to effort. The use of VMI in SAP APO™ (cf. Section 4.2.3.B) might be considered as an alternative.

4.4   Service Parts Planning with SAP APO™ and SAP ICH™

Compared with the primary products, service parts have some specific features that affect SCM. Some of these are a very high number of stock keeping units (SKU), stocking decisions for parts at locations, sporadic demand, and frequent supersession (replacement of one part by another part, perhaps because of technical or legal changes). Service parts are usually quite profitable. This motivated SAP (in alliance with Caterpillar Logistics Services and the Ford Motor Company) to develop a solution with a dedicated focus on service parts and their specifics. According to the nature of the development partners, the main industry focus of the first release SAP SCM™ 5.0 is engineering, construction, and automotive. For the next releases it is planned to extend this focus to other industries.

The Service Parts Planning (SPP) solution follows the "one size fits all"-approach – no ABC analysis is used for planning. Instead there is a high degree of automation – e.g., checking whether the current forecast model is appropriate and an automatic forecast model selection – and control mechanisms as procurement approval. A limitation of the SPP solution is that currently there is no link to the installed base in order to derive the demand for service parts (e.g., gear belts for automotives) based on the operating time or the mileage, and preventive maintenance is also not considered in this solution. Depending on the value of the service part, repairing is an alternative to procurement. This is especially the case in the aerospace industry. A common concept is to have remanufactured parts – e.g., engines – in a stocking and repair cycle. The planned availability of the remanufactured service part depends on the number of "old" parts and the repair capacity. Within the current release of service parts planning only very basic functionality is offered for this process.

SPP is not a separate application system but is based on new and disjunctive functionality, mainly in the SAP APO™. The complete solution for Service Parts Management also contains Service Parts Execution (SPE) in SAP ERP™, SAP CRM™, and in SAP EWM™. SPE contains, e.g., the following processes:

- Sales, claims and returns, and entitlement management based on SAP CRM™, the ATP check in SAP APO™, and the delivery processing in SAP ERP™. The order fulfillment processes include sales from stock and sales from a third-party vendor to the customer.
- Procurement execution, including ASN creation in SAP ICH™ and validation in SAP ERP™.
- Stock transfer execution in SAP ERP™.
- Warehouse management in SAP EWM™.

(A) Process Overview of Service Parts Planning

The purpose of service parts planning is ensuring that the service parts are available at the right location in order to meet the target service level. Fig. 4.32 shows the processes that SPP offers for this purpose.

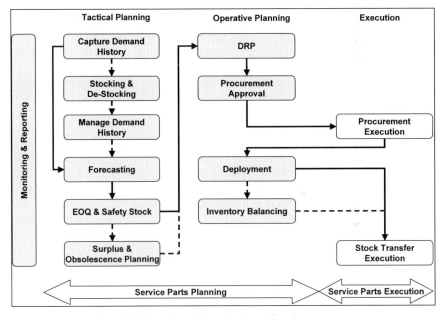

Fig. 4.32 Overview of Service Parts Planning processes

The processes for tactical planning span the medium- to long-term range and are based on time periods of fiscal years, months, or even weeks. The processes starting from the capturing of historical demand to the stocking decision, the adjustment of the demand history, the forecasting, the determination of the economic order quantity (EOQ) and of the safety stock levels (and optionally surplus and obsolescence planning) are considered as tactical. The results of the tactical planning are

- decisions about stocking or de-stocking,
- forecast based on the demand history, and
- safety stock and economic order quantity

for each location product and, optionally, the scrapping of surplus quantity. This information is used by the operative planning processes which are based on daily periods.

Distribution Requirements Planning (DRP) determines the required procurement quantity depending on the stock and order situation in the network. The procurement proposals are checked in the procurement approval process and released. Sending the releases to the supplier, receiving Advanced Shipping Notifications (ASN) from the supplier, and posting goods receipt are parts of the procurement execution. Based on the received goods the service parts are replenished to the target locations in the deployment process. Inventory Balancing enables lateral stock transfers between locations in exceptional cases in order to avoid shortages. In both cases stock transfer orders are created and sent to SAP ERP™ for execution. The result of service parts planning is that the location facing the customer – the delivery plant – should have sufficient inventory to cover the customer's requirements.

In-house production is not considered in this solution; if the company produces the service parts itself, the production is considered in the same way as that obtained from an external supplier.

(B) Service Parts Planning and SAP APO™

Though SPP covers similar processes to the components DP and SNP in SAP APO™, it relies on completely new functions. Compared with DP and SNP, SPP offers some additional functions, lacks some functions (e.g., aggregated planning and macros in forecasting), and uses a different logic for similar functions. Another aspect is that SPP does not offer much flexibility to model processes in different ways by re-arranging the process flow across the components of the application system. There is also no process interface with the "normal" SAP APO™ functions, and mixing SPP with DP, SNP, or PP/DS for the same location product is not intended and might result in inconsistencies.

(C) Service Parts Supply Chain

One of the specifics of the service parts solution is that the supply network has a tree-like structure with one or more entry locations (this is where the supplier delivers to) and for each entry location (optionally) one or more child locations. Looking from the demand side, a strict single sourcing policy is assumed. This fixed and hierarchical distribution structure is modeled as a Bill of Distribution (BOD). Fig. 4.33 shows an example.

The BOD is used throughout the whole service parts planning solution – from capturing demand to inventory balancing. There is a functional separation between locations that deliver to the customer (customer-facing locations) and locations that deliver to other locations (parent locations). If one location does both, SPP assigns the customer-related data (e.g., forecast, safety stock, customer orders) to a virtual child location for this location, and the original location keeps only the transactional data for its role as a parent location. This implies that a stock transfer is modeled from the location to its own virtual child location in order to cover the customer-related demand.

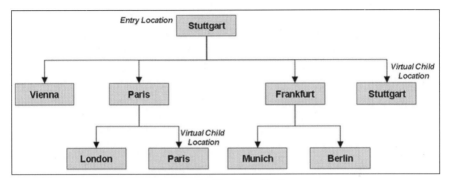

Fig. 4.33 Bill of Distribution

Restrictions in the modeling of the BOD are that the tree structure must be respected and that a location can only be used once within one BOD. It is, however, possible to have multiple entry locations within one BOD – e.g., to model different supply networks in Europe and America. Fig. 4.34 shows some BOD structures.

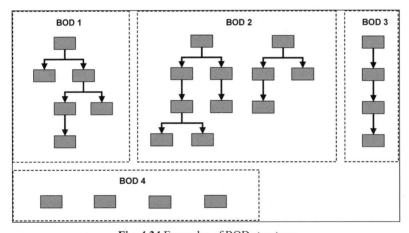

Fig. 4.34 Examples of BOD structures

In the SPP solution the customer-facing location and the first stockholding location are distinguished in order to keep the demand history consistent throughout changes of the stocking decision. The customer-facing location is the first location which is checked. However, it might not be able to confirm the customer request, either because it was decided that the service part should not be stored at this location or because it has run out of stock. Since rules-based ATP (see Section 4.2.5.B) is used for the service parts management solution, other locations are checked and the sales order will probably be confirmed in one or more of these other locations. The locations which confirm the customer request are the ship-from locations. The first stockholding location, however, is the first location in the check sequence that is defined as stocked (i.e., the replenishment indicator is set to "stocked") and should therefore be able to confirm the customer request.

The demand history for SPP is based on the first stockholding location. If the customer-facing location and the first stockholding location differ, a check is made during upload of whether an ATP rule with a location substitution procedure for both locations exists.

Locations that are geographically close and have only a small demand can be grouped to a virtual location for consolidated ordering (VLCO). This way the VLCO is considered as one location in DRP planning, and all transactional data (stock, demand, and fixed receipts) is aggregated to one location. As a consequence, netting is performed within the locations of the virtual consolidation location and only orders for the net demand are created. The distribution to the real locations is done in deployment and inventory balancing.

Another specific of the service parts supply chain structure is the inclusion of contract packagers, who perform packaging and re-packaging steps for the ware-houses within the BOD as subcontractors. These packaging steps might be required for any location within the BOD. The "normal" subcontracting process has the downside that the information on the goods movements at the subcontractor's ware-house is not visible. Therefore, the contract packagers are modeled as a special type of MRP area within the location they are assigned to. The contract packager is obliged to perform the goods receipts and goods issues with the SAP EWM™ system, and therefore the inventory and the goods movements of the contract packager are integrated into service parts planning.

4.4.1 Forecasting

Since service parts planning is mainly a make-to-stock or a procure-to-stock process, planning relies almost entirely on forecasts. The forecast is calculated on the basis of the customer demand history using statistical forecast models. In a similar way to that followed in DP (see Section 4.2.1), forecasting is applied within the boundaries of the company and does not support any collaboration with the customers (Fig. 4.35).

One of the most specific characteristics of service parts is the high proportion of products for which demand is sporadic. Most service parts are required if there is a failure in the primary product – due to wear, accident, or other events. These failures are hardly predictable, and there is a multitude of potential causes for failure. Failures caused by a specific service part are usually comparatively rare, which results in sporadic demand. However, there are also fast movers among service parts.

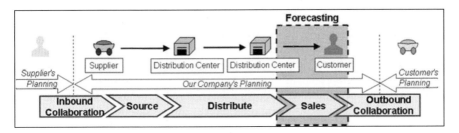

Fig. 4.35 Scope of forecasting within Service Parts Planning

The forecast drives the procurement and replenishment of service parts either directly or indirectly as an input for safety stock determination. Fig. 4.36 shows an overview of the forecasting process as part of SPP, including the necessary process of capturing and managing the demand history.

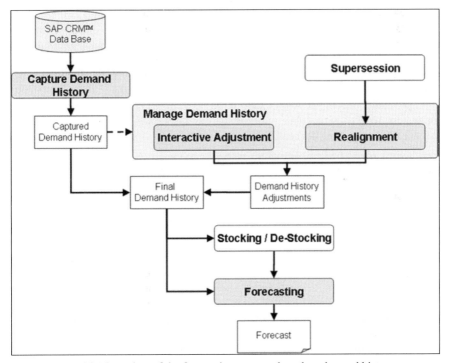

Fig. 4.36 Overview of the forecasting process, based on demand history

(A) Capture and Management of Demand History
The demand history is loaded from SAP CRM™. During the upload the data is processed to fit the specifics of SPP, e.g., aggregating the demand along the BOD and per period, scaling the demand history to the number of workdays (to exclude the impact of the calendar on forecasting), and determination of the facing and the first stockholding location. As an optional step it is possible to adjust the demand

history – examples are realignments due to changes in stocking decisions, super-session, or separating demand caused by promotions. Another example of demand adjustment is removing demand caused by a unique and exceptional occurrence (e.g., replacement of exhaust pipes due to a legal change) because this increase in the demand is not representative and should therefore not influence future fore-casting. The final demand history contains the captured demand and the adjustments and is used for stocking decisions and for forecasting.

(B) Forecasting
The forecast model and its parameters have a significant impact on the quality of the forecast. Since service parts planning usually deals with a huge number of service parts, the forecast model and its parameters are adjusted automatically in the process steps "model evaluation," "model selection," and "parameter tuning." If the performance of the forecast model [measured by Trigg's tracking signal (Dickersbach 2007)] exceeds the defined thresholds and the settings allow changes of the forecast strategy, an automatic model selection is performed and the forecast parameters are tuned. These parameters are used for forecasting. The combination of these four process steps (model evaluation, model selection, parameter tuning, and forecasting) is called composite forecasting. The next process steps within forecasting are disaggregation, calculation of the standard deviation, and release of the forecast. Fig. 4.37 shows the process steps within forecasting.

If applicable, tripping is used in addition, so as to measure whether the smoothing parameters are still appropriate, as explained by Dickersbach (2007). If this is not the case the forecast model is reinitialized.

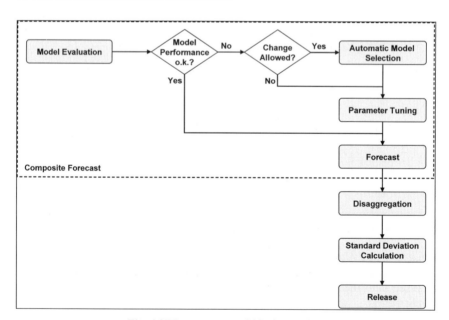

Fig. 4.37 Process steps within forecasting

Forecasting is performed in the planning book at location and product level. It is not possible, however, to use macros or aggregation for planning (see Section 4.2.1.B).

The forecast models in SPP are univariate models – in the current release no causal relationships can be considered. Compared with DP in SAP APO™ there are some additional forecast models – especially the intermittent model for slow-moving parts (Dickersbach 2007) – and some are lacking, e.g., multiple linear regression. Depending on the forecast model, outlier correction is inherent, possible, or impossible. The forecast is calculated for the demand quantity, the number of order items, and the average demand per order item. Of these three parameters only two are calculated independently, and the third is computed from the other two parameters.

As Fig. 4.38 shows, the forecast is calculated on two levels, on a detailed level (for each location product at the child locations) and additionally on an aggregated level (at the parent locations). The forecast at the aggregated level relates to an aggregation along the BOD and not to an aggregation of products – which is not possible in SPP. The reason for the aggregation along the BOD is that the forecast becomes less sporadic and that the forecast is usually more accurate for a larger data base. In the next step the forecast on aggregated level is disaggregated according to the detailed forecasts of the individual child locations.

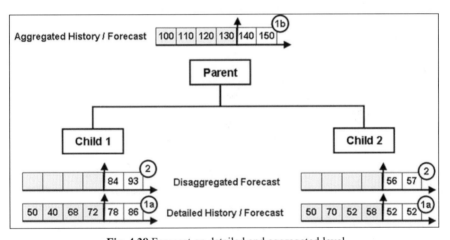

Fig. 4.38 Forecast on detailed and aggregated level

The forecast is first performed at the detailed level for each location product based on the demand history (1a). Within the same forecasting run the forecast is also performed at the aggregated level (1b), based on the aggregated history along the BOD. In a second step the aggregated forecast is disaggregated dependent on the values of the detailed forecast (2). The values obtained will typically differ from those computed in the detailed forecast.

The forecast error is calculated as a standard deviation between the forecasts and the demand history. Forecasting in SPP does not use the ex-post forecast but the forecast history, i.e., the forecasts that were calculated in the past (differing from DP; see Section 4.2.1.A).

The forecast is released automatically as long as the deviation of the new forecast to the previous forecast does not exceed the threshold defined in the forecast profile. Otherwise the forecast has to be released interactively. Only released forecasts are used by DRP (see Section 4.4.3).

(C) Product Life Cycle Planning
Planning of the phase-in and phase-out of service parts is based on the demand history of reference products (*cf. desirable feature 2.3.3.1.c*). Both phase-in and phase-out forecasting for new products is done in two steps. In the case of phase-in forecasting, for example, in the first step the demand history is scaled and stored in the phase-in profile, and in the second step the scaled demand curve of the phase-in profile is applied for the new product. The procedure differs from that used in DP (see Section 4.2.1).

4.4.2   Inventory Planning

Inventory Planning is concerned with determining the safety stock levels in the internal supply chain. The first task for Inventory Planning is deciding whether a distribution center (DC) should be stocked with a service part at all, and, if true, the second task is to determine the required amount of safety stock. Since the safety stock also depends on the order quantity, the economic order quantity (EOQ) is calculated in inventory planning as well. These tasks are focused on the internal supply chain as visualized in Fig. 4.39.

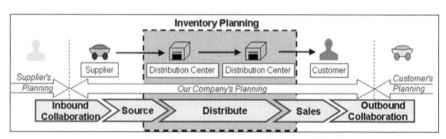

Fig. 4.39 Scope of inventory planning within Service Parts Planning

Inventory Planning contains two separate processes (Fig. 4.40): the stocking and de-stocking decision and the calculation of safety stock and the EOQ. Usually, the stocking and de-stocking decisions are taken before forecasting and the safety stock and EOQ calculations are performed after forecasting.

(A) Stocking and De-Stocking
One characteristic of service parts planning is the huge number of products and locations. At the same time, many of the service parts are slow movers. Therefore

the decision of whether to keep a service part in stock at different warehouses is more significant than for regular products. In order to reduce capital lockup and warehouse costs, not all service parts are kept at all locations. The decision on whether a product is stored at a location depends on the demand for and the costs of the product and is controlled in SPP by the replenishment indicator (a parameter in the location product master data) (*cf. desirable feature 2.3.3.1.j*).

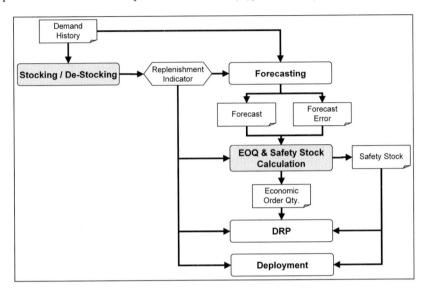

Fig. 4.40 Stocking decision process overview

The stocking and de-stocking planning runs determine the replenishment indicator based on rules which are defined per location. An example for such a rule is shown in Fig. 4.41, where the procurement costs and the past demand quantity define the thresholds for stocking or de-stocking.

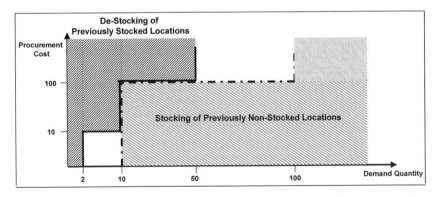

Fig. 4.41 Decision logic for stocking and de-stocking

Other criteria could be the number of order items or the product group. The rules for stocking and de-stocking are defined per location. It is possible to define stability periods per product to avoid an oscillation between stocking and de-stocking. The location-specific list of stocked products and locations is called the authorized stocking list. There is, however, no standard report to display this result for multiple products.

(B) EOQ and Safety Stock Calculation

Safety stock levels are especially important for the service parts, because there is usually an immediate demand for them; any service part required should be available from stock, and its non-availability could be disastrous. On the other hand, the sum of the safety stocks might become very high because of the large number of SKUs.

The purpose of safety stock planning is to determine the safety stock levels required to meet the target service levels (TSL) (*cf. desirable feature 2.3.3.1.r*). The safety stock depends on the uncertainty of the lead time and of the demand during the lead time, and on the order quantity. To take the impact of the order quantity into account, the safety stock and the EOQ are calculated together. The EOQ calculation is based on the trade-off between order costs and stockholding costs. Like all costs in SAP APO™, these cannot be transferred from costing or finance applications.

EOQ and safety stock are used by DRP to determine the requirements for procurement. Deployment also considers the safety stock for calculation of the replenishment quantities and EOQ for rounding.

4.4.3   Distribution Requirements Planning and Procurement

The purpose of Distribution Requirements Planning (DRP) is to calculate the net requirements for the whole internal supply network (i.e., the BOD) and create procurement proposals. DRP covers the net demand of the child locations by a planned stock transfer from the parent location, which results in a demand at the parent location – shifted by the lead time. The net requirement of the parent location is covered by a planned stock transfer from the next level (the parent location's parent location), and so on until the entry location is reached. The net requirement of the entry location is covered by a procurement proposal – either a schedule line or a purchase requisition. Fig. 4.42 shows the demand and the material flow for a two-level BOD.

The DRP plan is displayed in a planning book called "DRP matrix." It is not possible to plan interactively in the DRP matrix. It is, however, possible to create additional fixed receipts and requirements.

DRP planning creates, changes, and deletes planned stock transfers and schedule lines (or purchase requisitions). Depending on how far in the future the order changes are due, not all changes suggested are desired. It is possible to restrict the number of changes per planning horizon.

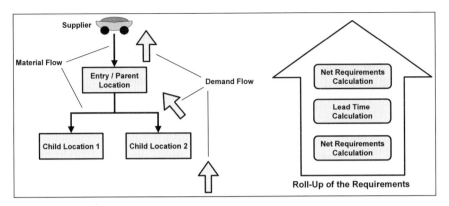

Fig. 4.42 Planning tasks in Distribution Requirements Planning

Forecasting is usually carried out with a granularity of months or fiscal years, whereas for DRP the granularity is finer, based on data relating to single days. The forecast is disaggregated evenly into the daily periods. Therefore, changes in the forecast – e.g., because of seasonality – may lead to sudden changes of the daily forecast at the borders of the forecast periods. These leaps can be smoothed using the functionality for anticipated demand coverage, where a portion of the demand difference is already considered in advance, as shown in Fig. 4.43.

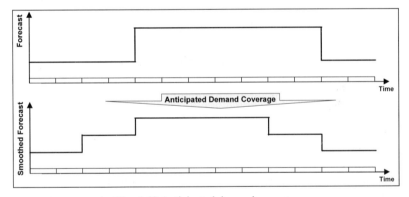

Fig. 4.43 Anticipated demand coverage

If a product is subject to a seasonal demand, the safety stock might also change from season to season. If the seasonal behavior differs, e.g., because winter starts earlier, the planned safety stock for a period might not be sufficient. Therefore it is possible to apply the maximum safety stock to the next few periods and thus to increase the safety stock.

Since most of the service parts are slow movers, DRP offers an option for product group procurement in order to reduce the fixed order costs. The assumption is that a part of the fixed order costs results from a set-up at the supplier's plant, and therefore products with similar set-up procedures are grouped together. Product group procurement, however, is only carried out if the savings in terms of the fixed order costs exceed the costs caused by holding stock. To determine the optimal interval for product group procurement, the total costs are calculated for all alternatives from monthly to yearly (increasing by monthly steps). Another option offered by DRP is the explicit modeling of shutdown times at the supplier's plants, e.g., during summer vacation. In this case the demand is sourced earlier. Fig. 4.44 shows the process overview of DRP and Procurement Approval.

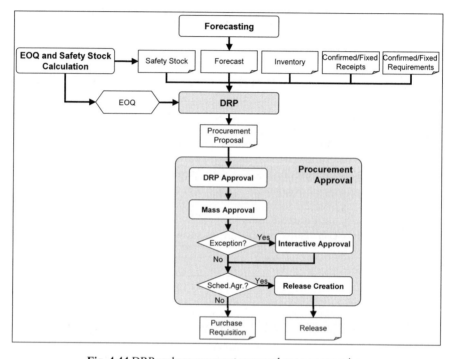

Fig. 4.44 DRP and procurement approval process overview

(A) Procurement Approval

The results of the DRP process are proposals for external procurement – either schedule lines for scheduling agreements or purchase requisitions for contracts – which have to be approved before release of the procurement approval and its dispatch to the supplier. The purpose of the procurement approval process is to prevent an unusually high procurement – quantity or value – without notice. To this end, up to three different approval steps are performed:

- The approval service (which might be integrated with DRP) checks whether any rules are violated and creates alerts.
- Mass approval allows checking of the value (the total value and the difference from the last DRP run) for selected products and approval or rejection of the whole selection.
- Interactive approval is required for those scheduling agreement items or purchase requisitions that create alerts.

The procedure for procurement approval is shown in Fig. 4.44. Releases are only created after procurement is approved. The basis for the procurement approval is the approval rules – for example, if the ordered value exceeds a threshold. Procurement approval is followed by the procurement execution, where the information is sent to the supplier. It is always the complete delivery schedule that is approved, and not the individual schedule line (see Section 4.2.3.A). Depending on compliance with the rules and the approval step, the status of the scheduling agreement item changes.

(B) Procurement Execution
For execution of the procurement, the release is sent to the supplier. The common method is to use EDI for large suppliers and those with a well-developed IT landscape. For small and medium suppliers it is, however, possible to apply SAP ICH™ to integrate them into the planning environment for service parts management with comparatively little effort. Fig. 4.45 shows the document flow for procurement.

It is essential for this process that the supplier sends ASNs to SAP ERP™. The release is sent from SAP APO™ to SAP ICH™. In SAP ICH™ the supplier creates an ASN which is sent to SAP ERP™ as an inbound delivery. In SAP ERP™ the ASN is validated by a plausibility check (e.g., whether a scheduling agreement item for this product with the supplier exists). After the validation an acknowledgement is sent back to SAP ICH™ and the inbound delivery is sent to SAP EWM™ as an inbound delivery notification (IDN). This inbound delivery notification is activated automatically to become an inbound delivery in SAP EWM™, and goods receipt can be posted.

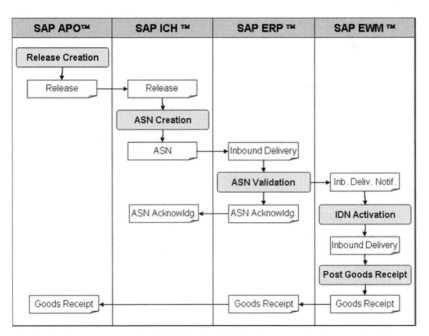

Fig. 4.45 Process steps and document flow for procurement execution

4.4.4 Deployment and Inventory Balancing

Deployment and Inventory Balancing are concerned with the replenishment of the child locations of the BOD (Fig. 4.46). While DRP calculates the requirements along the BOD in order to determine the procurement quantity at the entry location, Deployment creates stock transfer orders from the parent location to its child locations. There might, however, be cases where a demand at a child location is not covered by its parent location but there is excess at another child location. Inventory Balancing checks whether this is the case and creates lateral stock transfers between child locations or to parent locations from different branches of the BOD (or even from a different BOD).

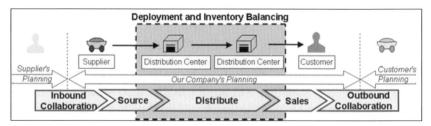

Fig. 4.46 Role of deployment and inventory balancing within Service Parts Planning

(A) Deployment
The stock transfers calculated by Deployment are always due today, i.e., no stock transfer orders are created for future dates. Another feature of Deployment is that the type of the demand is grouped into "tiers" (such as backorders, forecast and stock transfers, and safety stock) and the priority of the tiers is taken into consideration for the determination of the deployment quantities. Deployment in SPP differs not only in this respect from Deployment in SNP (see Section 4.2.2.I). Deployment does not rely on the results of a previous DRP run, but recalculates the requirements in the part of the BOD using the DRP algorithms. Therefore, Deployment is possible even without planned stock transfer requirements.

Deployment is performed from the parent location to its direct child locations. If the BOD contains more than two levels, multiple Deployment runs are required. For each parent location a Deployment run is required to replenish the inventory from the entry location to the customer-facing location (Fig. 4.47).

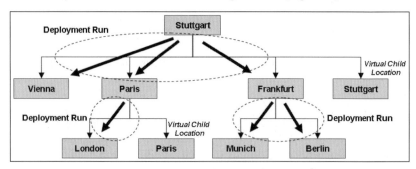

Fig. 4.47 Deployment runs

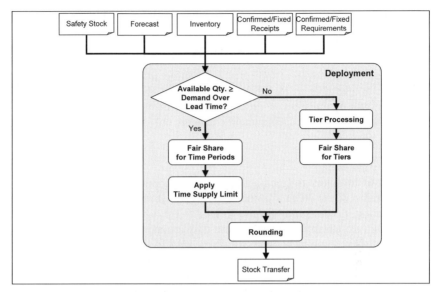

Fig. 4.48 Deployment process overview

Depending on whether or not the available quantity for deployment (ATD, Available-to-Deploy) at the parent location exceeds the demand of the child locations over the lead time, deployment proceeds in two different ways as shown in Fig. 4.48.

If the quantity available is greater than the demand over the lead time, a fair share calculation is performed to determine the deployment quantities for each location (*cf. desirable features 2.3.3.1.t and 2.3.6.3.e*). These quantities might be reduced if they exceed the accepted storage quantity of the target location. The accepted storage quantity is represented by the time supply limit (TSL). Rounding is performed as a last step.

In the other case the parent location does not have enough quantity available to cover the children's requirements over the lead time. The demand over the lead time at the child location is an overdue demand at the parent location. For this overdue demand tier processing is performed, which means that the demand elements are sorted according to their priority tiers (*cf. desirable feature 2.3.6.1.a*). The purpose of assigning different demands types to priority tiers is to ensure that the most important demand is covered first by Deployment when the available quantity is less than the demand over lead time. In the subsequent step the fair share calculation is applied for the demands of the tier with the lowest priority in order to determine the deployment quantities, and rounding is performed. In the fair share calculation the demand of the tier that cannot be fully covered is the only one considered.

There are two different ways to perform Deployment in SPP: pull deployment and push deployment. Pull deployment is performed by regular deployment planning runs and is triggered by a need at the child location – in a similar way to the Deployment in SNP (see Section 4.2.2.I). In contrast to this, push deployment is triggered by a goods receipt at the parent location, and the goods are not stored in the warehouse but sent to the child locations. Push deployment is intended for fast-moving parts and has the advantage that it leads to faster replenishment along the BOD. Whether pull or push deployment lead times are used for planning is controlled by the deployment indicator – a setting in the product master data. Deployment is also able to expedite a shipment by choosing a faster but more expensive means of transport in order to cover a demand in time (*cf. desirable feature 2.3.6.2.f*). However, Deployment considers only the transport relationships of the BOD.

(B) Inventory Balancing
The purpose of Inventory Balancing is to supply locations laterally – i.e., not along the regular paths of the BOD – in the case of need at one location and surplus at another location (*cf. desirable features 2.3.4.2.c, 2.3.6.2.e, and 2.3.6.2.h*). Lateral stock transfers are usually avoided because they are more costly: The service part has to be transferred one more time, which means additional costs, and the lateral transfer itself might be less cost efficient owing to low volume. In exceptional cases a lateral stock transfer is nevertheless necessary. The nature of Inventory Balancing is short term and is intended in most cases as an optional step if deployment fails to cover the demand. Inventory Balancing might also be triggered by a de-stocking decision or by supersession in order to get rid of the predecessor product. An example of Inventory Balancing is shown for a BOD with a very simple structure in Fig. 4.49.

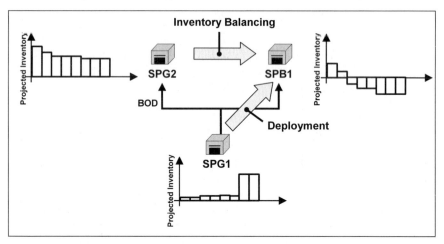

Fig. 4.49 Example of inventory balancing

The BOD contains the parent, SPG1, and two children, SPG2 and SPB1. The normal route of supply is via Deployment from the parent to the children along the BOD. In this example, however, there is a shortage at the child SPB1, which cannot be covered by the parent. Therefore the shortage at SPB1 will persist even after deployment. Inventory Balancing now checks whether there is another location close by that has a surplus. Whether a location is considered to be "nearby" is modeled by the inventory balancing area. If this is the case, a transportation lane exists between the locations, and the savings that would accrue from a stock transfer exceed the costs by a threshold, a lateral stock transfer is created – in this example from SPG2 to SPB1. Inventory balancing only creates stock transfers which do not follow the BOD relationship.

The algorithm for Inventory Balancing starts with determining locations with need and locations with excess, and it ensures that neither the excess inventory nor the need is just a momentary state and that the parent location will not cover the need before the lateral stock transfer arrives. Inventory Balancing creates stock transfers only when the difference between the savings and the moving costs exceeds a predefined threshold. The following savings are considered:

- Inventory savings as a proportion of the deployed quantity of the annual inventory costs.
- Warehouse space savings represent the benefit that follows freeing up warehouse capacity.
- Service benefit takes the prevented loss of an order into account.

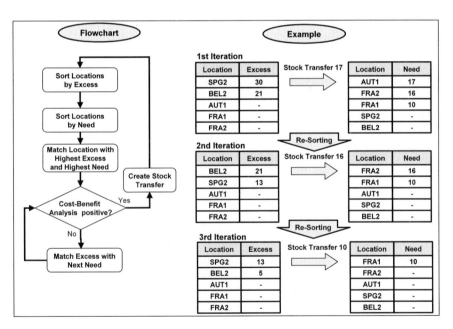

Fig. 4.50 Flowchart and example for matching of demand and excess

The inventory and the warehouse space savings apply at the source location and the service benefit at the target location. The moving costs consist of the goods issue costs at the source location, the transport costs, and the goods receipt costs at the target location. Fig. 4.50 shows the flowchart and an example.

The excess and the need are determined for all locations within the inventory balancing area. The locations are sorted by their excess and their need, and the highest excess is matched with the highest need. If the cost-benefit analysis shows a favorable situation, a stock transfer is created; otherwise the excess is matched with the next need, and a new cost-benefit analysis is performed. After each stock transfer the locations are re-sorted by their excess and need. This flow is repeated until all excess locations have been processed. The stock transfer orders may be released automatically or require interactive approval.

4.4.5   Surplus and Obsolescence Planning

The goal of surplus and obsolescence planning is to identify inventory within the internal supply chain which exceeds the projected demand and therefore only consumes warehouse space. In order to identify the surplus and remove it from the warehouses concerned, first the total surplus within the supply network is determined. This determination is based on the total expected demand (including a safety buffer), the total available stock, and the stock in transit. In a second step the total surplus is disaggregated – in other words, the surplus quantities for the individual locations are determined. If the value of the surplus is within the predefined limits, orders for scrapping the surplus are created automatically. In the other case the scrap orders need to be approved interactively. Fig. 4.51 gives an overview of the surplus and obsolescence planning process.

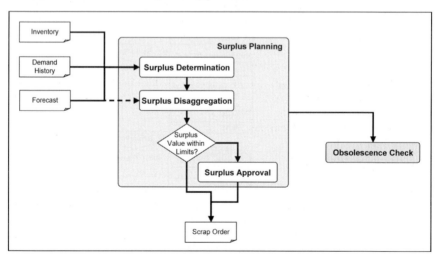

Fig. 4.51 Surplus planning and obsolescence check process overview

The procedure for surplus determination differs depending on whether the service part belongs to a primary product that is still produced or whether the primary product is already out of production. If the primary product for the service part is no longer produced and the retention date for this product is reached, this product will become obsolete once the stock is used up. In this case the parameter "flag as obsolete" is set per location product by the surplus planning service. The obsolescence check service monitors whether there is no more stock or stock in transit and whether all location products are marked as obsolete. In this case the parameter "obsolete" is set at product level and the service part can be removed from the system.

The service part is considered as a "past model part" if the primary product that requires this service part is no longer produced. This is the case if the field "production end date" in the product master data is set and the year (and not just the date) has passed. If the service part is a past model part, the demand is calculated using phase-out forecasting (see Section 4.4.1.C). The end of the service part's planned availability is marked by the end of the retention period, which is calculated in a separate step (Dickersbach 2007). The retention period describes the horizon for the availability of a service part after the production end date of its primary product. Even after production of the primary product has ceased, there may be legal requirements or contractual obligations to keep the service part available for a mandatory retention period. Often the service part is kept available for additional periods. Only if the demand forecast falls below a predefined threshold, it no longer makes economic sense to keep the service part available.

The surplus is determined for the whole BOD or for the sub-BOD below the entry point. In order to remove the surplus, it is necessary to determine the surplus per location. This is called surplus disaggregation. The surplus disaggregation starts with determining the recommended quantity to be kept at each location and determines the surplus quantity per location on this basis. It always starts with locations that are not stocked. For these the retention quantity is zero. The stocked locations are sorted in descending order according to the demand history, the forecast, the potential surplus quantity, or the potential surplus in number of days. Sorting of the locations by ascending demand history implies that the location with the lowest demand history is the first to remove its surplus. If the value of the surplus order at a location is above the threshold for the surplus approval, an interactive surplus approval is required.

4.4.6 Supersession

Replacement of a service part with another service part (or with several other service parts) is called supersession. Reasons for supersession might be changes in the product, changes in the service parts portfolio, or changes in the sales package – e.g., from a kit to separate service parts or the vice versa. The supersession functionality in SPP relies on calculated dates for the using-up of the predecessor and the planning start date for the successor service part. The key dates for supersession are:

- The pending obsolescence date (POD) is the earliest date that the successor product can be used.
- The successor product planning date (SPPD) defines when planning for the successor product starts.
- The successor product receipt date (SPRD) is the first date on which the successor product is available at the entry location.
- The stock exhaustion date (SED) is the projected stock-out date for the predecessor.

The POD is maintained as a master data setting, and the other dates are calculated only after the POD is reached. Fig. 4.52 visualizes the interdependencies between these dates.

The stock exhaustion date (SED) is the date by which the stock for the predecessor product is used up. This date is used as the basis for calculation of the successor product planning date (SPPD) as the date on which the planning for the successor product has to start. The scheduling elements for this calculation are the (longest) lead time from the supplier to the entry location plus the lead time within the BOD plus optionally a buffer. SED, SPRD, and SPPD are updated daily.

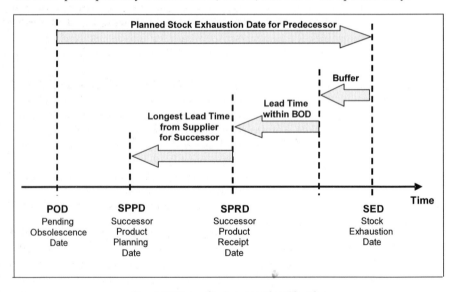

Fig. 4.52 Dates for Supersession Planning

Based on these dates the SPP solution considers supersession explicitly in the processes "Capture and Manage Demand" and DRP (Fig. 4.53).

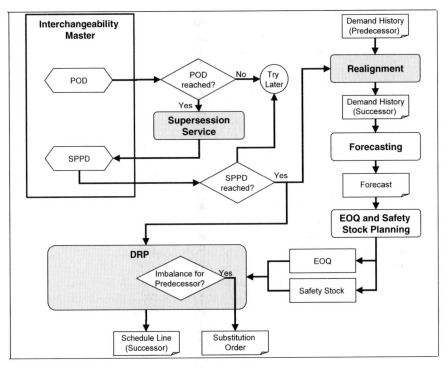

Fig. 4.53 Service Parts Planning with supersession

If the SPPD is reached, the history of the predecessor is copied to the successor part. Forecasting, EOQ, and Safety Stock Planning are able to handle the successor part in the same way as any other part. If everything runs smoothly, the demand for the predecessor is covered from the remaining stock and new supplies are already available to cover the demand for the successor. In case of an imbalance – either because the stock for the predecessor is not sufficient to cover the demand (for the predecessor) or because there is stock left – DRP creates substitution orders to cover the demand for the successor with supply elements of the predecessor or vice versa.

Deployment does not include the substitution orders in the deployable quantity, so that excess supplies of the predecessor at the entry location will not be transferred to cover a demand for the successor product at the child locations. This will cause an imbalance in short-term planning, which will be resolved by the rules-based ATP check for the sales order.

4.5 Forecasting and Replenishment with SAP F&R™

The purpose of SAP F&R™ is planning the short-term replenishment of a retailer's stores and distribution centers. Retailers usually have a very high number of SKUs, often tens of millions. This implies the necessity of grouping products and mass maintenance of master data. Other retailer specifics are the integration of Point of Sales (POS) data and the multitude of demand influencing factors (DIF), e.g., holidays, sport events, and weather conditions. On the other hand, availability of the articles at the source (distribution center or supplier) is usually no constraint, and the focus of planning is only the next order. Though the processes in SAP F&R™ are similar to those in DP and SNP, the automated processing of this huge amount of data with retail-specific requirements justifies a separate application system. SAP F&R™ does not use the liveCache but the database structures for time series (TSDM) and orders (ODM).

SAP F&R™ is a planning system and can be integrated with SAP ERP™ for retail or with legacy systems. Assuming that SAP F&R™ is integrated with SAP ERP™ for retail, the master data – products, locations (stores, distribution centers, and suppliers), and transportation lanes – and the transactional data (POS, inventory, and open orders) are transferred via programs obtained from SAP ERP™. The master data objects are the same as in SAP APO™, but SAP F&R™ offers special transactions for mass maintenance.

During automatic replenishment planning SAP F&R™ performs a forecast on article and location level, calculates the net demand for the location, and schedules it. Forecasting takes the demand influencing factors automatically into account; they have to be maintained by the planner before. The automatic replenishment results in order proposals, which are transferred to SAP ERP™ either as stock transfer orders (from DC to store or from DC to DC) or as purchase orders (from DC to store or from supplier to store). It is possible to process the order proposals before they are sent to SAP ERP™, and exceptions supports the planner in selecting the critical order proposals. Fig. 4.54 provides an overview of the process based on the assumption that a promotion is planned.

Forecasting and replenishment planning is performed within the process step "automatic replenishment." The following steps are executed in the background:

- Automatic adjustment of the history (POS data, stock transfer orders) by outlier control and stock-out correction.
- Classification of the articles according to their sales frequency into six classes.
- Forecasting of future demand based on the POS data (for stores) or the stock transfer orders (for distribution centers). During forecasting the effects of the demand influencing factors in the past and their impact on the future are considered. The forecast is performed for each article at each location with weekly granularity. Different forecasting algorithms are used depending on the sales frequency classification.
- Based on the target service level and the forecast error, a safety demand is added to the forecast.
- The weekly forecast is disaggregated according to a profile. The disaggregation is either static or dynamic – in the latter case the pattern for the

disaggregation might be adjusted (but the weekly quantity remains fixed). Procurement constraints, e.g., such as replenishment on Tuesdays and Thursdays only, are taken into account.

- The net requirement is calculated as the sum of the receipts (inventory plus fixed receipts) minus the forecast, minus the presentation stock (for stores), and minus the confirmed stock transfer demands (for DC).
- The net requirement is rounded to package sizes.
- Comparison of the fixed ordering costs with the capital lockup by inventory might result in a change of the order size; the bigger the fixed ordering costs and the lower the capital lockup, the bigger the proposed order quantity will be.
- Restrictions are taken into consideration.
- Exceptions are created for unusual order proposals such as an unusual quantity or a new product.

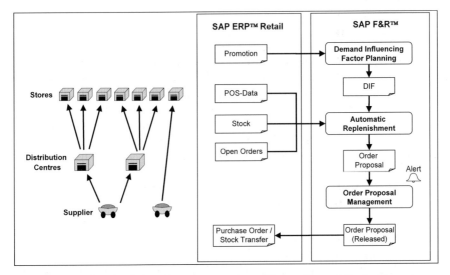

Fig. 4.54 Supply chain and process overview of Forecasting and Replenishment

The result of this procedure is the generation of order proposals and exceptions – if an order proposal exceeds quantity or value limits, for example. The order proposals are either automatically released or, if they caused any exceptions, interactively checked in the order proposal management. The order proposal is converted into a stock transfer or into a purchase order, depending on the source.

This procedure is performed for each location – both for the DC and for the store. Forecasting at the DC is necessary because the stock transfers from the stores are only calculated for the very short term, which is not sufficient for replenishment planning of the DC.

4.6   Controlling and Support Processes with SAP SCM™

4.6.1   Tracking and Tracing with SAP EM™

The purpose of SAP EM™ is to monitor supply chain processes, especially beyond the borders of the own company. For example, the external procurement of electronic parts for a European company from the Far East might involve multiple parties – the supplier, a carrier in Asia, an Ocean carrier, and a carrier in Europe – and include the steps shown in Fig. 4.55.

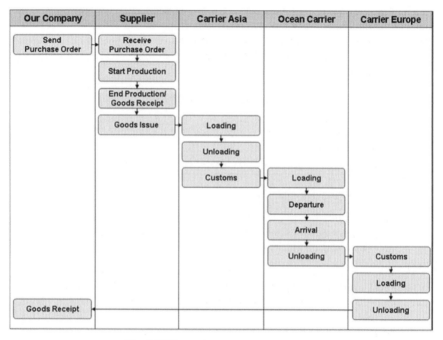

Fig. 4.55 Events in a procurement process

Most of these steps take place outside the boundaries of the company regarded. If an exception occurs – e.g., if the carrier in Asia missed the ship – it is very valuable to know this as soon as possible. The sooner this information is available, the more time is left to look for alternatives.

In SAP EM™ this process is modeled by an event handler that contains the steps shown in Fig. 4.55 as expected events (*cf. desirable feature 2.3.8.1.a*). For each purchase order an event handler is created automatically. The events are related to the purchase order, and if the event for loading the ship is not recorded within the allowed time window, the planner is notified. In this way it is easily possible to identify the purchase orders concerned.

An event is regular if it is reported within the expected time window, and early or late if it is reported outside this time window. If the event is not reported within a specified time frame it is regarded as unreported. There are also unexpected events – for example, breakdown of the supplier's production lines, breakdown of a truck, or missing documents at the customs – that might have an impact on the schedule. For any of these cases it is possible to define rules to change the status of other events, to create and filter alerts, or to adjust data (*cf. desirable feature 2.3.8.1.b*).

It is also possible to monitor supply chain processes for more complex order chains in which more than one order is involved. An example of this is the order fulfillment where sales order, delivery, and shipment need to be monitored. In this case each order relates to an event handler, and the different event handlers are linked by an event handler set corresponding to the order fulfillment process.

Events can be reported from multiple sources. For SAP ERP™ there are pre-configured interfaces. If events are transmitted via EDI, XML, SMS, or by means of RFID (see Section 4.7.2), the integration might require further configuration. Another option for reporting an event is to use the Internet. SAP EM™ offers a Web user interface where supply chain partners can log on and maintain a record of the events. For each partner, the authorization concept restricts the view on the data to the section that is relevant for it (*cf. desirable feature 2.3.8.1.c*).

Examples for the use of SAP EM™ are

- transportation by sea, including customs processing,
- fulfillment of make-to-order including production, delivery, and billing, and
- procurement, including transport and invoice.

Mini Case: A very striking example is the use of SAP EM™ for monitoring of ocean-freight logistics: soon after the implementation of EM hurricanes caused havoc at two US ports. With the visibility provided by SAP EM™ it was possible to reroute the shipments to other ports and thus prevent damage to the shipments (Dießner and Rosemann 2007).

4.6.2 Alert Monitoring

The concept of the alert monitoring is to notify of critical situations that require attention from or intervention by the planner (*cf. desirable feature 2.3.8.2.a*). Since most of the planning tasks are usually performed automatically as batch jobs and the amount of planning data is usually huge, exception reporting by alerts is one of the key features of SAP SCM™.

Alert monitoring is available for all the SAP SCM™ application systems, and there are different alert types per component. Two different kinds of alert monitors are used, one for SAP APO™ (also used by SAP F&R™) and the other for SAP ICH™

(which is also used for SPP). They differ slightly in the way alerts are handled – i.e., in the display, sorting, confirming, and forwarding of alerts. An example for an alert is a shortage situation resulting from a production shortfall (cf. *desirable feature 2.3.5.3.c*).

Alerts are either displayed in the alert monitor as a stand-alone application or integrated into other planning functionalities. The alert types are either calculated via macro functionality (see Section 4.2.1.B) or pre-defined. The relevant objects (e.g., products or resources) and alert types can be defined individually for each planner.

A feature of the alert monitor in SAP APO™ that supports the planner in resolving the alert situations is the option to freeze the alerts. In this case any change in the alerts is marked – which is especially helpful if the alerts are created during a simulation. This way it is possible to experiment with actions and see their impact on the alert situation. Depending on this impact it is possible to decide whether to save the actions or not. In the SAP APO™ alert monitor it is also possible to branch from the alert into the functionality required to resolve the alert (e.g., into a production planning function in the case of shortage). However, this functionality depends on the alert type and does not always provide all the desired options.

Because alerts are performance critical, it is possible to calculate them in the background – e.g., after the batch jobs for planning have finished – in order to provide the planner with the relevant alerts in the morning. It is also possible to send certain types of alerts regularly per batch job to the SAP office or to another specified e-mail address.

An advantage of the alert monitor in SAP ICH™ is that it also notifies the supplier and allows exchange of information on the alerts.

(A) Supply Chain Cockpit
In SAP APO™ the supply chain cockpit provides an overview about the supply chain structure, both graphical and as a list by element (locations, products, resources, and other). The benefit of the supply chain cockpit is that it links the supply chain structure with an overview of the alerts (cf. *desirable feature 2.3.1.a*). The entries in the bottom row display whether there are current alerts for any application and priority. Another functionality of the supply chain cockpit is to execute queries for planning data in SAP APO™ or for data from SAP BI™.

(B) Monitoring for Service Parts Planning
For SPP there are two additional monitors in SAP ICH™: the shortage monitor and the SPP cockpit. The shortage monitor displays the criticality (based on the days' supply), the open quantity, and the stock on hand for service parts. The shortage monitor also allows the supplier to have a look at the problems that may be relevant for it. The SPP cockpit, on the other hand, provides an overview of the planning situation of a service part within the BOD – e.g., with respect to stock, stock in transit, and forecast – even if there is no exception in planning.

4.7 Data Integration and Master Data

4.7.1 System Landscape and Data Integration

Though it is possible to use SAP SCM™ systems in conjunction with legacy systems, the typical case is integration with one or more SAP ERP™ systems. Fig. 4.56 shows the application systems within SAP SCM™ (with the exception of SPP), their integration into SAP ERP™, and the technology used for integration.

The integration between SAP APO™ and SAP ERP™ is performed using the core interface (CIF), which is based on queued remote function calls (qRFC), applies business-related logic for data consistency, and offers functionality for monitoring and administration of the interface. Each creation, change, or deletion of an order triggers a data transfer, so that data between SAP ERP™ and SAP APO™ is tightly integrated. The integration of data history from SAP ERP™ to the SAP BI™ structures within SAP APO™ relies on the info structures of the logistics information system (LIS) in SAP ERP™. The general idea is that SAP APO™ is the master for planning and scheduling and SAP ERP™ for execution.

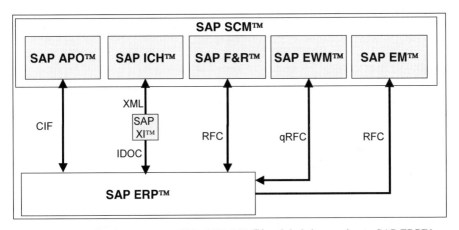

Fig. 4.56 Application systems within SAP SCM™ and their integration to SAP ERP™

SAP ICH™ is connected with SAP ERP™ via the Exchange Infrastructure (SAP XI™) based on XML messages on the SAP ICH™ side and IDOCs on the SAP ERP™ side. The integration of SAP F&R™ with SAP ERP™ requires the retail solution on the SAP ERP™ side and is based on RFC calls. SAP EWM™ receives inbound or outbound deliveries from SAP ERP™ and sends goods receipt or goods issues back via qRFC. SAP EM™ receives data from SAP ERP™ via RFC.

The application systems within SAP SCM™ use separate transactional data – e.g., the inventory in SAP APO™ is not visible in SAP ICH™. Any user company requiring integration between these application systems needs to set this up specifically itself. To some extent, however, there is a common use of master data (e.g., the product master data; see also Section 4.7.3).

The system landscape for Service Parts Management (SPM) contains SAP ERP™, SAP CRM™, SAP APO™, SAP ICH™, SAP EWM™, and SAP Netweaver™, including SAP XI™ and SAP BI™. Data exchange between SAP CRM™ and SAP APO™ is required for sales history (SPP) and for the ATP check (SPE). Though some of the functionality might also work in a different system landscape, e.g., with SAP APO™ and SAP ERP™ alone, so far SPM is released only for this system landscape. Therefore SPP (including monitoring of service parts) is also dependent on this system landscape (Fig. 4.57).

The relationship between the application systems must not necessarily be one to one. For example, it is quite common to connect multiple SAP ERP™ systems to one SAP APO™ system. Connecting one SAP ERP™ system to multiple SAP APO™ systems is less common, and it depends on the business processes whether it is possible for the same location products. For SAP EWM™, on the other hand, it is likely that there will be separate systems for each warehouse and therefore an *n* to 1 relationship to SAP ERP™.

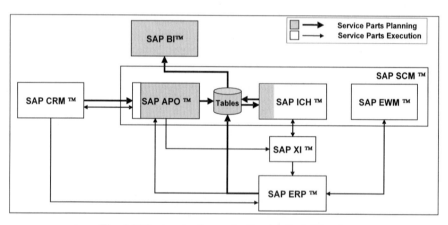

Fig. 4.57 System landscape for Service Parts Planning

4.7.2   RFID with SAP Auto-ID Infrastructure

The software to adapt the RFID information to the planning and execution systems of SAP is called Auto-ID Infrastructure (SAP AII™). It bridges the information acquisition by RFID readers and the usage of RFID-generated data in business processes. Data is filtered, processed, and provided to related ERP systems almost in real-time, whereas related ERP data is upgraded with respect to the RFID functionality needed. Regarding movements of goods, for instance, SAP AII™ facilitates a near-real-time view of the real world events in the ERP system. SAP Netweaver™ technology is used by SAP AII™ for integration with other systems such as SAP ERP™, SAP SCM™, SAP CRM™, and also non-SAP systems. EPCglobal standards, e.g., for Application Level Events, are supported.

The architecture of the SAP AII™ solution is shown in Fig. 4.58. The RFID tags are connected to fixed or mobile devices by the use of electromagnetic or inductive coupling. The various devices are managed by device management software such as noFilis CrossTalk. The device management is the interface between RFID devices and SAP AII™ and provides it with a standardized interface to the various RFID devices.

The device management software and SAP AII™ communicate via proprietary XML or the standardized Physical Markup Language (PML). All incoming messages are handled by the SAP AII™ message input. For testing purposes, a SAP Auto-ID Test Tool can simulate RFID-read events and pass them to the message input. All messages are stacked up before being processed by the message mapping and the workflow-based rule engine, which interconnects RFID raw data with defined business processes. The results of the rule engine are predefined activities which are triggered by SAP AII™. The interface with other systems, such as SAP ERP™, third-party ERP systems, or the SAP Object Event Repository™ (SAP OER™) can be conducted by SAP Exchange Infrastructure (SAP XI™), as shown in Fig. 4.58. Another option is the direct connection of SAP AII™ and these systems via IDoc, XML, RFC, or Web Service. The usage of SAP XI™ is beneficial for RFID deployments that affect several information systems.

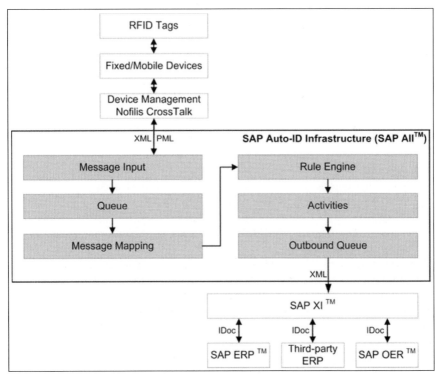

Fig. 4.58 Simplified architecture of SAP AII™ and surrounding systems

SAP AII™ contains a set of preconfigured templates with corresponding rules and activities. These templates are currently available for the business processes "Picking and Goods Issue," "Goods Receipt," "Returnable Transport Items," "Kanban," and "Warehouse Management." These RFID-enabling templates are briefly described below.

(A) Picking and Goods Issue
After receiving a purchase order from the buyer, the vendor picks or produces goods and builds a handling unit (HU). A HU consists of the goods that are to be shipped and attached packing instructions. The HU is then created in the SAP ERP™ system. The Electronic Product Codes (EPC) of the items included are associated with the HU. During goods issue, the EPCs are scanned and a verification of the purchase order fulfillment is executed. Finally, an Advanced Shipping Notification (ASN) with the loaded goods is created and sent to the buyer.

(B) Goods Receipt
The ASN of the vendor is used by the buyer's SAP ERP™ system to create an event handler in SAP EM™. In addition, the expected EPCs are registered with SAP AII™. During goods receipt, the EPCs are scanned and a consistency check is conducted by SAP AII™. In addition, SAP AII™ posts the goods receipt event to SAP EM™ SAP ERP™.

(C) Returnable Transport Items
Returnable Transport Items (RTI) are means to assemble goods for transportation, storage, handling, and product protection in the supply chain which are returned for further usage. The SAP AII™ template contains the identification of the incoming and outgoing RTI and their tracking and tracing. The tracking and tracing process, which is conducted in cooperation with SAP EM™, contains the state of an RTI, e.g., "full," "empty," or "damaged," and RTI-related reporting functionality.

(D) Kanban
Kanban-enabled production is widely adopted as a concept of lean management. SAP AII™ supports RFID-enabled kanban, with RFID gates sensing the kanban cards. Replenishment is then triggered by SAP AII™ in SAP ERP™.

(E) Warehouse Management
The warehouse management integration facilitates transparency concerning movements of stock and goods. Processes at the warehouse can be coordinated better because of the higher visibility of events taking place in the real world. The location of goods, for example, is known nearly in real-time, while manual scanning is avoided.

4.7.3   Master Data

Master data has an important role in SAP SCM™. It defines the supply network and controls many processes (*cf. desirable features 2.3.1.b and 2.3.1.k*). Most of the master data objects, e.g., the material master, are transferred from SAP ERP™ via CIF (see Section 4.7.1). Since SAP SCM™ offers additional planning functions compared with SAP ERP™, the master data contains additional parameters in SAP SCM™. These parameters can be maintained in SAP SCM™, but are usually set or derived during the transfer by customer-specific enhancements in order to simplify master data maintenance. Table 4.3 provides an overview of the most important master data objects in SAP SCM™ and their correspondence in SAP ERP™.

If multiple SAP ERP™ systems are connected to one SAP SCM™ system, harmonized master data is a precondition for comprehensive planning processes.

Table 4.3 Corresponding master data objects in SAP SCM™ and SAP ERP™

SAP SCM™ Master Data	SAP ERP™ Master Data
Location	Plant, Supplier, Customer
Product	Material
Resource	Work Centre (PP) or Resource (PP-PI)
Transportation Lane	Purchasing Info Record
Production Data Structure (PDS)	Production Version
Production Process Model (PPM)	Production Version
Procurement Relationship	Info Record, Scheduling Agreement, Contract

The most complex master data objects are the production data structure (PDS) and the production process model (PPM), which contain the information on BOM and routings. Within these objects scrap is considered (*cf. desirable feature 2.3.5.3.d*) and joint production can be modeled (*cf. desirable feature 2.3.3.1.n*). PDS and PPM are used alternatively and correspond to the production version (a combination of BOM and routing/recipe); in order to transfer master data for production planning to SAP APO™, maintenance of production versions in SAP ERP™ is required. The PPM is the more mature master data, but will not be developed any further. The PDS offers some advantages over the PPM – mainly the integration of engineering change management (ECM) and object dependencies – but does not yet cover the entire functionality of the PPM.

Another important master data object is the resource. It models the capacity for production, and it is possible to model calendars and different shift sequences for each resource (*cf. desirable feature 2.3.5.2.d*).

The substitution of products, for example at the end of their life cycle, is defined in the interchangeability group (*cf. desirable feature 2.3.1.h*). This relationship is considered in planning.

Organizational entities, such as company codes or sales organizations, which are very significant in SAP ERP™, are not used in SAP SCM™.

For a broader discussion of SCM master data we refer to Wood (2007).

Chapter 5

Supply Chain Management Case Studies

5.1 Case Study Colgate-Palmolive

5.1.1 Company Profile

Colgate-Palmolive Company is a global provider of consumer products that make lives healthier and more enjoyable in more than 200 countries and territories. The company focuses on strong global brands in its core businesses – oral care, personal care, home care, and pet nutrition. With ~70% of sales derived from international operations, Colgate-Palmolive maintains strong global growth. About 10 million tubes of toothpaste are sold worldwide every day.

The company is headquartered in New York and has about 36,000 employees. Its main research center is located in Piscataway, NJ. In 2007 Colgate made almost 14 billion USD in sales.

5.1.2 Objective of the Use of SAP APO™

The consumer goods industry is struggling with volatile demand resulting from the buying behavior of the retailers. Intensive advertising campaigns cause peaks for the provision of supplies. Further demand peaks result from the buying habits of the end-consumers at the weekend (e.g., bulk buying in supermarkets) and after the payment of wages and salaries.

To master these demand peaks, the retailers set a high value on short lead-times for their orders from the suppliers. These in turn are anxious, particularly in the case of their large customers, to match and integrate the logistics carefully.

Because of the characteristically high standard of cleaning required for the production facilities and the long setup times these entail, the lot size optimization assumes particular importance in this type of consumer goods industry.

In general, the production lines should not be set up during a shift. This results in lot sizes being a multiple of the production capacity of a shift, and the lot size planning must be carefully matched with overall logistics.

5.1.3   Parties Involved in the "Extended Enterprise"

It is obvious that companies will endeavor to use CPFR® (cf. Section 2.1.2.4) to solve the aforementioned problems. In this connection, the Colgate-Palmolive group wishes to stock its customers actively at the retailing level. They use VMI (cf. Section 2.1.2.8) to collaborate with large customers in order to react efficiently and replenish more efficiently.

It is planned that a cooperative SCM with the suppliers will be handled via the Inventory Collaboration Hub (ICH) in the future. Closely related are the so-called co-packers. They add value to the Colgate-Palmolive products by, for example, producing special packaging in the form of vacation or promotion sets.

The independent transport companies do not represent any significant constraints and thus are not considered in the APO model.

5.1.4   Structure/Network

Approximately 85% of Colgate's factories and its distribution warehouse volume are planned in SAP APO™. VMI is practiced with 175 customer distribution centers, which are also integrated in the supply network, and this number is increasing. Several thousand stock-keeping units (SKU) are planned with APO. They represent individual articles in their positions, in particular also in warehouses that the supplier maintains in customer plants.

As Fig. 5.1 illustrates, Colgate-Palmolive is primarily concerned to perform SCM in close cooperation with the retailers and their distribution centers. This "philosophy" is reflected in the organizational structure by the fact that "Customer

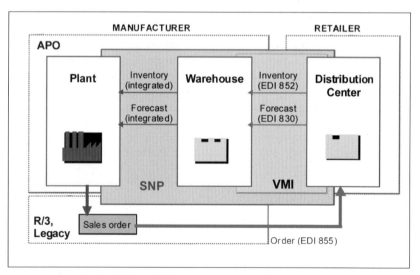

Fig. 5.1 SNP and VMI processing at Colgate-Palmolive

Account Teams" plan the demand and the deliveries together with the customers. The VMI implementations (cf. Section 2.1.2.8) take account of the considerably fluctuating retail demand. The EDI procedure forms the technological basis, using the ANSI X12 exchange format with the message types 852 (Inventory), 830 (Forecast), and 855 (Order).

5.1.5   IT Infrastructure

Colgate-Palmolive has been a user of SAP systems for many years and operates extensive SAP ERP™ and SCM installations. In 2007, more than 99.6% of the business activities were supported by SAP systems. One sales region, e.g., Europe or Asia, is assigned to one ERP System and one SCM System. Each of these systems runs on its own computer, although the computers are all physically installed at the same location (in the Colgate-Palmolive Data Center in New Jersey).

An intensive data exchange takes place, primarily with the customer's distribution centers, either daily or weekly. The incoming data must be checked very carefully. For this reason, data transfer at short time intervals is not considered desirable.

The Internet is currently being used for communication with the suppliers (Supplier Managed Inventory); in future it will also be used in conjunction with the planned CPFR.

5.1.6   SAP APO™

Colgate-Palmolive was one of the first large pilot users of the APO. SAP provided special support for the project because it is representative not only for the consumer goods industry but also for other industries. Executives at Colgate-Palmolive have expressed their satisfaction with the SAP solutions publicly many times.

The group communicated its requirements to SAP in mid-1997. The pilot software was supplied in July 1998. The project team began its work at the same time. APO became productive in the United States in April 1999.

The Colgate-Palmolive group currently uses SCM in its North America, Latin America, Europe, and Asia Pacific divisions. The following SCM modules are used:

- APO Demand Planning: The elements used by Colgate-Palmolive are Statistical Forecasting, Phase-in/Phase-out, and Like Modeling.
- APO SNP: The particular elements used by Colgate-Palmolive are the VMI, DRP, Deployment, and Transport Load Builder subsystems (cf. Section 4.2.2).
- APO PP/DS: Colgate-Palmolive utilizes the repetitive manufacturing functionality in PP/DS. In particular the PP/DS Optimizer for setup optimization, Multi Resource Planning Heuristic, Product Planning Table, and alerts.

- APO TP/VS: The elements used by Colgate-Palmolive are route and load optimization, carrier selection, and load tendering.
- APO Global ATP: This functionality is only utilized in the United States. GATP supports the real time inventory availability checking for customer orders and back order processing business scenarios. It is important for Colgate-Palmolive to use ATP, particularly for the so-called backorder processing for customer orders that have already arrived: The customer orders contain priorities that need to be confirmed and acted upon. If necessary, deliveries already confirmed to the customers must be modified.
- Supply Chain Event Manager: Colgate-Palmolive uses this functionality to track the status of customer shipments.
- Inventory Collaboration Hub (ICH): The particular element used by Colgate-Palmolive is the Responsive Replenishment functionality. Colgate is utilizing this functionality to support their VMI process for both direct (customer) and indirect (distributors) scenarios. The ICH functionality will eventually replace SNP VMI as the preferred tool to support Colgate's VMI processes.

Colgate-Palmolive plans a long-term coexistence of SAP's SCM solutions and the ERP Systems. The latter handle the execution of the plans, e.g., printing of production schedules, delivery notes, and invoices. The close integration within the SAP systems simplifies the reduction of the lead times for order processing with regard to the goals and success factors mentioned above. APO provides Colgate-Palmolive with the functionality needed to deliver the right products to the right locations at the right time. The project was realized in several stages.

Stage One:
  APO SNP/VMI Pilot in the USA
  Replaces Manugistics DRP, Deployment and Colgate-Palmolive developed, VMI customer data translator and order generator.
Stage Two:
  APO Demand Planning Pilot in the USA
  Replaces R/3 Flexible Planning functionality.
Stage Three:
  APO TP/VS Pilot in the USA
  Replaces Colgate-Palmolive's legacy transportation planning system.
Stage Four:
  APO PP/DS Pilot in the USA
  Replaces R/3 PP Repetitive Manufacturing functionality.
Stage Five:
  Accelerated Roll-out of APO Functionality Globally
  Implement APO solution in Europe, Asia, and Latin America.

5.1.7   Use of Additional SAP Products

Colgate-Palmolive is a strategic SAP customer and has implemented much of the SAP functionality. Some of the major functionalities include the Business Information Warehouse, Supplier Relationship Management, Trade Promotion Management, Customer Relationship Management, and Human Resources.

5.1.8   Interview

with Jim Newkirk, Director, Global Supply Chain & Technology Development, Colgate-Palmolive

CV

Jim Newkirk is responsible for the configuration and development of Global Supply Chain and Technology IT applications utilizing SAP's suite of software products, including Advanced Planner and Optimizer, Enterprise Buyer, Business Warehouse, Product Lifecycle Management, and R/3.

Jim started his career with Colgate in 1991 as a co-op programmer in the Jeffersonville development group. In 1992 he became a full-time Colgate employee working as a junior programmer in the areas of manufacturing and sales order management applications development. At the start of the Focus 125 project in 1994, Jim was named team leader for the co-development and implementation of SAP's R/3 Production Planning module. In 1998 he joined the SAP APO Supply Network Planning implementation team as the team leader of the Vendor Managed Inventory (VMI) module. In 2000 Jim relocated to Morristown to join the newly formed Global Supply Chain Development team. He became the Global Supply Chain Development team leader for APO's Transportation Planning/Vehicle Scheduling and Production Planning/Detailed Scheduling modules. In 2004, Jim's area of responsibility was extended to include the management of the entire Global Supply Chain Planning Systems team. Jim earned his Bachelor of Science degree in Computer Science from Indiana University.

Alexander Zeier: Do you implement a more central or a decentralized SCM approach?

Jim Newkirk: When Colgate-Palmolive implemented APO back in 1999 we had regional supply chains providing products to our customers. Given that, we chose to take a decentralized approach by implementing an SCM instance for each of our regional ERP instances. Our business has evolved over the years and our supply chain is now more global. In order to better support our global supply chain we are now considering consolidating these SCM instances into one global instance.

Alexander Zeier: What are the key benefits and the most valuable KPI improvements for your company by implementing the SCM solution?

Jim Newkirk: Colgate uses a very homogenous SAP solution. What is the main value from the customer's perspective of using such a homogenous solution instead of using a diversified one? Colgate-Palmolive's IT strategy is to deliver the same systems, the same information, with the same performance, everywhere in the Colgate world. Our partnership with SAP and the use of SAP software has enabled us to realize this strategy. We are able to deliver standard SAP solutions to our business users that deliver the same information with the same performance globally.

Alexander Zeier: From your perspective, which is the area of the SAP SCM™ solution that still has the greatest potential for improvement?

Jim Newkirk: In my opinion the APO PP/DS functionality has room for improvement especially in the area of managing resource networks. Resource networks are the multiple levels of constraints that include storage tanks and routes for supplying products from making to finishing. These complex resource networks must be taken into consideration during detailed scheduling in order to create an executable production plan.

5.2 Case Study Danfoss

5.2.1 Company Profile

The Danfoss Group is a global company with net sales of about 2.7 billion Euro and approximately 22,000 employees worldwide. One of its business areas is Danfoss Drives, a leading global supplier for electronic motor controls. Its headquarters and one of its plants are situated in Grasten, Denmark. Other main factories are located in the USA, Germany, and China. Drives are used in very different areas, e.g. heating and ventilation, water applications, and food and beverages. Consequently the required properties of the drives vary, and Danfoss offers mass customization for its drives – based on a limited number of components and options, millions of variations and solutions are possible. These are assembled, tested, and delivered anywhere in the world within 48 hours.

5.2.2   SCM Based on SAP Systems

Associated with this high standard of service is the necessity for reliable confirmation dates to be supplied to the customer. Therefore the availability of the components and the capacity for assembling have to be checked in this assemble-to-order environment. Danfoss had developed this functionality, which is vital for its business, by itself as an addition to its earlier use of a SAP R/2 system. When switching from SAP R/2 to SAP R/3™, it had to be decided whether to develop this functionality anew in R/3 or to use packaged business software. In discussion with SAP it was decided to use the functionality for multi-level ATP in SAP APO™ for this purpose.

Only the assembly groups of the first BOM level – the basic units – are checked for availability. For the capacity check a workaround is used: A dummy component is added to the BOM, and for this dummy component an allocation check is performed. The allocation quantity (maintained in SAP APO™) represents the capacity in this case. Another specific feature of the implementation is that only the BOM information is transferred from SAP R/3™ (and not the production version containing BOM and routing), because SAP APO™ is only used for the multi-level ATP and not for production planning and scheduling. The duration of the implementation project (SAP R/3™ and SAP APO™) took about 15 months.

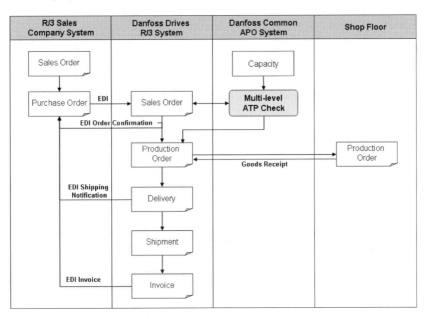

Fig 5.2 Sales Order Flow at Danfoss Drives

Fig. 5.2 shows the sales order flow across the Danfoss application systems. The sales order is centrally received in the Danfoss Sales Company SAP R/3™ system and transferred to the Danfoss Drives SAP R/3™ system. Here the sales order is checked using multi-level ATP functionality and the allocation check for capacity of SAP APO™. The confirmation is transferred back to the Danfoss Sales Company followed by the standard logistics execution processes.

This process using multi-level ATP and allocation check is implemented in the two plants in Denmark and in the USA. The implementation between these plants is aligned.

5.2.3 Interview

with Dan Bruun, Danfoss APO Information Architect, Kristian Klit, Danfoss Drives IT Director, Susanne Schurter, Danfoss Drives SAP Business Consultant

CV

Dan Bruun joined Danfoss in 2001 after receiving a master's degree in translation and interpretation from the Aarhus School of Business. He has worked within the area of SAP/ERP systems since 2002 first as a PP super user and joined Danfoss IT as an SCM consultant in 2005. As of 2007 he has been the APO information architect working with IT Governance and Quality of system solutions.

CV

Kristian Klit joined Danfoss A/S in 1977 and was a member of the first SAP project (SAP R/2) which was implemented in Danfoss Drives A/S in 1989. Since then he has worked as a super user in the logistics department until 1995 where he joined the Danfoss Drives A/S local IT department. In 2003 when Danfoss Drives A/S implemented the SAP R/3 system together with SAP SCM (APO) he was the project manager. Since 2004 Kristian has been IT Director at Danfoss Drives A/S.

CV

Susanne Schurter joined Danfoss A/S in 1981 and was a member of the first SAP project (SAP R/2) which was implemented in Danfoss Drives A/S in 1989. Since then she has worked as super user in the purchasing department until 1998 where she joined the Danfoss Drives A/S local IT department. In 2003 when Danfoss Drives A/S implemented the SAP R/3™ system together with SAP SCM (APO) she was a member of the project management.

Jörg Dickersbach: Did the implementation of SAP APO™ provide any additional advantages compared with the SAP R/2 development by Danfoss?

Kristian Klit and Susanne Schurter: We cannot say that this set-up has been a big business improvement – it has mostly been a 1:1 implementation compared with the R/2 set-up, so the service level to the customers has not changed very much and the smoothing out of the production load has not changed either.

But we must say that in the long run implementing SAP APO™ was the correct decision. Otherwise it would be difficult to get the benefits of new developments from SAP.

Small improvements of the SAP APO™ implementation are a reduction of workflow issues and the fact that the production for incomplete orders can already start and thus reduce the delay to the customer.

Jörg Dickersbach: What were the major obstacles in implementing APO?

Kristian Klit and Susanne Schurter: We needed a long time to get to know the complicated functionality of SAP APO™ and to fine-tune our developments. Now, more than 3 years after the implementation we feel that we have a stabilized system which in most cases fulfills our needs. Nevertheless, there are still improvements required, especially in the time conversion between R/3 and APO and between summer- and wintertime.

Major disadvantages of using APO are that it is not easy to find skilled APO resources and that very good business knowledge is required when problems occur. From a functional point of view many problems occurred because the times in R/3 and APO are not well aligned. Another problem is that at Danfoss Drives

JIT production is "real" JIT production, where the production of the goods starts in due time before the goods have to be shipped (requested delivery date – pick pack time – lead time). In APO, JIT was defined as production some time before the requested delivery time and then storage of the material until the pick pack date has been reached. Also, some functionality seems not to be finished (alert monitor as an example), so we still have to consider APO as a "new" system. Finally, from a system operation point of view, the support package test and the refresh of quality systems are very resource demanding.

Jörg Dickersbach: Are there any plans to extend the usage of SAP APO™, e.g., for production planning of the basic units?

Dan Bruun: Currently Danfoss is using the SCM functionality for Demand Planning, GATP, and Cross-System Flow of Goods. Within a short time we will also use PP/DS with the focus on CTP for stock transport requisitions and Product Allocation. Finally, we are currently running SAP Event Manager as a pilot project and doing a clarification project on SAP ICH™.

5.3   Case Study Henkel

5.3.1   Company Profile

The Henkel Group has its headquarters in Dusseldorf, Germany. The group is divided into three strategic areas of competence: laundry & home care; personal care; and adhesives, sealants, and surface treatment. In 2007 sales were more than 13 billion Euros (~20 billion USD). Henkel owns subsidiaries in more than 75 countries, and over 75% of Henkel's employees are located outside Germany. The firm's motto is "Henkel – A Brand like a Friend". This means that the company is particularly determined to quickly identify and fulfill consumer needs and to offer high service levels.

5.3.2   Objective of the Use of SAP Systems

For the past couple of years, the top management of Henkel has been paying more and more attention to the goal of reaching a high value of the firm, which means having a competitive Return on Capital Employed (ROCE). Of course that cannot be achieved only by increasing the numerator (profit) but also by decreasing the denominator (capital employed). IT has to make an important contribution to reaching this goal, and SAP Systems have a lot of potential to help.

With respect to SCM Henkel started with APO SNP because this module particularly influences the investments and thus the capital employed.

Up to now SAP products for SCM are heavily used on the sales side. Henkel plans to do some in-depth analysis on the supply side and is considering the implementation of more IT systems in this part of the supply chain.

The SCM concept at Henkel is based on SAP SCM™ 4.0 and was developed under the perspective of global usage. Applications started mainly within the European area; and Henkel is starting to refine the processes in Russia. APO is also running in the USA.

Some special challenges Henkel has to deal with are:

1. The customers (retailers) start more and more promotions, resulting in peaks in the sales curve. An example is an offer such as "Buy 3, get 4!"
2. The supplier's elasticity to react efficiently to peaks on the sales side of Henkel, since considerable lead times are caused by more customer-specific requirements.
3. Employees are challenged by the results of optimization algorithms that they have to rely on. The system analysts have therefore to proceed cautiously.

5.3.3 Parties Involved in the "Extended Enterprise"

There are several production plants serving several distribution centers (DC) belonging to the Henkel Group. These DC ship products to the DC of the customers, from where logistic service providers transport the products to the customer's stores.

5.3.4 IT Infrastructure

Henkel has two traditional R/3 platforms in Europe. APO runs on its own computer. APO's built-in data warehouse is used for reporting systems. For strategic planning the BW within SAP's BI product is used.

5.3.5 SAP Systems

5.3.5.1 SAP Systems in the Analysis Phase

When preparing the practical use Henkel made a very thorough analysis of the parameters that influence material management and production. The APO capabilities are used to simulate the supply chain using different assumptions and parameters.

The basis is the cornerstones concept. The so-called four-step cornerstones process was designed to be carried out every quarter.

Step 1 – The Segmentation Cornerstone

Henkel defined three quantity profiles:

- Standard quantities for items that are produced continuously
- Promotion quantities that are temporarily sold in larger quantities
- So-called exotic products that are made to order.

These profiles help to determine the "plannability" of sales.

Steps 2 and 3 – The Product Parameterization Cornerstone

APO is used

- for determining production frequencies and periods by leveling change-over and warehousing costs and smooth capacity utilization, and
- for determining optimal safety stocks to cover uncertainties in product supply and forecasting.

Step 4 – The Master Plan Cornerstone

The master plan has to ensure the reliability of the long-term production plans. This includes keeping to standard lot sizes, and adhering to cost effective production plans and detailed schedules. Once per quarter products are reclassified and the master plan for the next 3 months is determined. An inventory buffer counteracts the demand volatility and allows for continuous production of standard lot sizes.

Other components of the analysis are:

- Determining the minimum inventory level to guarantee the service levels.
- Intensive monitoring of the influences of service levels.
- Calculating the frequency and sequence of production processes so that production costs are minimized.
- Analysis of the portfolio to focus attention on products that are drivers of complexity.

An interesting property of Henkel's master plan is that technical equipment is not a "hard restriction" but can be modified, e.g., by replacing a large machine with two smaller ones in order to gain more flexibility though this means additional investment for equipment.

The analysis is repeated to take account of new developments in markets, distribution structures, and production technology. Christof Steinmeister, of Henkel's Central Material Management of the Laundry and Home care business section, calls it "dynamic analysis on interactive Key Performance Indicators (KPI)".

5.3.5.2 SAP Systems in the Operation Phase

The SNP functions are used to map the networks both of ordering and manufacturing locations and of contract manufacturers. Fig. 5.3 shows the supply chain planning process together with the SAP modules applied.

All sales organizations in the network enter their requirements online, using the collaborative planning capability of SAP APO™ together with the SAP Internet transaction server for those countries without direct access to APO via SAP GUI.

Sales forecasts are generated using DP. After adding special promotional quantities, the sales planning data is available. It is an input for requirements planning using the SNP capabilities of APO. Henkel creates and submits a rolling demand plan for the next 16 weeks. This data is taken to derive a detailed production plan for the first 4 weeks and to establish a rough capacity check for the outstanding 12 weeks. This guarantees that products are delivered to the distribution centers and subsidiaries on time.

The lot sizes and manufacturing periods that were determined by the segmentation and the master plan are also considered.

Detailed planning is carried out on a weekly basis using a 4-week horizon. It is automated using Henkel's special Process Flow Scheduler (PFS+). Equipment is allocated based on the master plan and the optimum frequency and production sequence. Resource planning also takes place at this level.

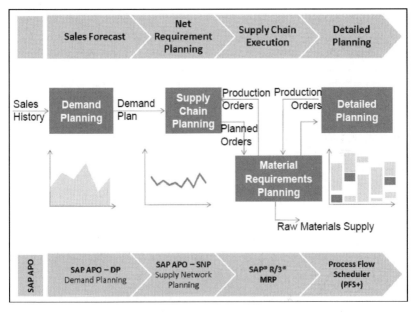

Fig. 5.3 Henkel's planning process

In the case of bottlenecks the Management by Exception (MbE) principle is applied. KPIs to plan and control the process are taken from the SAP BW™. Following the MbE philosophy, so-called Network Planner's Laundry Lists are created, e.g., if inventory is outside the inventory corridor. These lists also contain a special capacity check, which is used in rough planning.

The system generated quantitative improvements such as

- increases in full and on-time delivery service levels in the ranges of 97% and 100%, respectively, the average rose from 97.3% in 2005 to 98.7% in 2007, and the reach of the inventory even decreased slightly,
- a reduction in overall supply chain costs by 4%, and
- smoothed inventory levels.

5.3.6   Use of Additional SAP Products

For strategic and long range planning Henkel uses SAP SEM™ and the SAP BI™.

5.3.7   Interview

with Christof Steinmeister, Director Supply Chain Optimization, Henkel Laundry and Home Care Division

CV

Christof Steinmeister joined Henkel in 1991 as an electrical engineer. He was responsible for implementing SAP R/3 PP/PI within Henkel UC business. In 1998 Christof joined the Laundry and Home Care business as Materials Manager; in 2004 he took over the responsibility for Central Materials Management. He contributed substantially in the supply chain planning structure of Henkel's Laundry & Home Care business.

Peter Mertens: Can you give us some information on the accuracy of the forecasts within SAP's Demand Planning module?
Christof Steinmeister: With our monthly forecasts we reach about 80% in high-volume markets where the statistical basis is large enough.

Peter Mertens: When you used more sophisticated algorithms in Demand Planning, did they bring a higher accuracy than simple forecasting procedures?
Christof Steinmeister: We do not have enough experience since we are using only the traditional methods offered in SAP systems, such as exponential smoothing.

Peter Mertens: Do you experiment with RFID?
Christof Steinmeister: Yes we do, together with big customers like Metro or Wal-Mart.

Peter Mertens: Which new functionalities should SAP provide from your point of view?

Christof Steinmeister: We have some smaller problems with aggregating stock keeping units (SKU): One way is to aggregate stocks by product for forecasting purposes, independently of their location, the other one is to make aggregated forecasts per distribution center and then to break down the results to SKU.

5.4 Case Study Hilti

5.4.1 Company Profile

The Hilti Group is a world leader in development, manufacturing, and marketing of added-value, top-quality products for professional customers in the construction industry and in building maintenance.

The product range covers drilling and demolition, direct fastening, diamond and anchoring systems, firestop and foam systems, installation, measuring and screw fastening systems, and also cutting and sanding systems. Hilti is committed to excellence in innovation, total quality, direct customer relationships, and effective marketing. Hilti operates in over 120 countries around the world.

Worldwide, close to 20,000 employees work at headquarters and in sales organizations, engineering, customer service, production plants, and research and development centers. Two-thirds of the workforce has direct contacts with end-customers. Hilti's corporate headquarters (Hilti AG, abbreviated to HAG) is located at Schaan in the Principality of Liechtenstein.

One very specific feature of Hilti is its direct sales approach. About 15,000 final products are sold via the worldwide regional market organizations (MO), which typically are also legal entities. Owing to this vertically integrated organization, Hilti keeps in close contact with its customers. The typical structure for an MO contains one or two central warehouses (CW) and up to 200 Hilti Centers (HC) (cf. Fig. 5.4). Worldwide, more than 1,000 HCs serve the customers. In large countries such as the USA a third level with distribution centers (DC) is established in addition.

5.4.2 IT Organization and Evolution of IT Systems

The globalized IT, employing about 400 employees worldwide, is organized in three main strategic locations. Application development, engineering, European support and global group coordination are based at the company's headquarters. First- and second-level support is provided in Tulsa, USA, and Kuala Lumpur, Malaysia. About 40 people work in the Process Competence Center SCM, supporting primarily IT solutions for Manufacturing, Purchasing, Supply Chain Planning and Execution as well as Product Data Management.

Originally Hilti ran a wide range of different legacy and ERP systems at headquarters and in the market organizations. In the first wave of system standardization the marketing organizations were expected to use Oracle Applications as the ERP system and Manugistics SCPO as the Advanced Planning System. This package was labeled "H1", and the servers were placed locally. Manugistics was one of the early APS vendors and offered its systems under this name from 1992 to 2006, when it was acquired by the JDA Software Group.

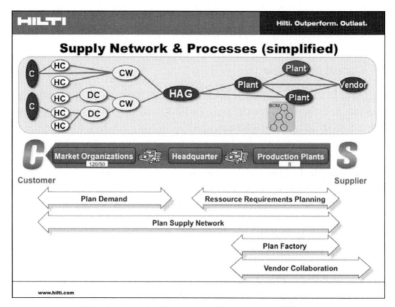

Fig. 5.4 Supply Network and Processes at Hilti

In 2000 the business-driven Hilti IT strategy was developed, and as a consequence the globalization of processes and data (GPD) was initiated. Consequently, in 2001, after a detailed evaluation, Hilti decided to realize a single system concept, based on SAP systems (R/3, APO, BW, etc.). The technical solution was named "H2". These two together formed the largest initiative Hilti had ever started: GPD/H2. This package brings standardized processes, data, and system solutions into the subsidiaries. The number of concurrent users of SAP R/3™ is up to about 5,000. As of today, data of 10 million customers are handled in the single R/3 system. The transaction volume is about 50,000 customer orders per day. The very successful rollout of R/3, APO, BW, CRM, and HR into the main MOs is already finished. Now even smaller organizations will be part of this initiative.

One global instance each of SAP R/3™, SAP SCM™, and the SAP Business Warehouse made up the core of the GPD/H2 technical solution. Meanwhile, other SAP solutions, such as PI/XI, EP, CRM, CUA, and Solution Manager, are also in use.

To build up its knowledge about SAP APO™ as soon as possible, Hilti developed interfaces between the existing SAP R/2 system operated at headquarters and APO in 2002, and set the PP/DS solutions of the mySAP SCM 3.0 system to become productive in 2003. In 2004 the SAP R/3™ system went live at headquarters and the SCM system was connected to R/3, followed by an upgrade to mySAP SCM 4.0.

In the next phases various ERP and legacy systems were replaced at all production facilities and marketing organizations by access to the central SAP R/3™ system and the central SCM system.

Early in 2006 the IT board took the strategic decision to switch to new x86 processor-based server technologies running on the Linux operating system. Consequently, in autumn 2006 the upgrade to SAP SCM™ 5.0 took place before the roll-out of the GPD/H2 planning solution SCM to the local marketing organizations was started.

5.4.3 Procedure Model for Conversion to SAP Systems

For the SAP APO™ roll-out into plants and the major marketing organizations a procedure model with about 200 activities has been defined. It is recommended that there should be at least a few weeks between switching to SAP R/3™ and to SAP APO™. This is based on the fact that APO is heavily dependent on stable master data from R/3. Thus, it may be necessary to introduce interfaces between SAP R/3™ and the Manugistics system for a limited time period.

Typically an SCM conversion project is finalized in about 3–4 months. Most of this time is needed to train the users in the new system and establish the processes. Stress tests play an important role during the conversion process. When Hilti is replacing Manugistics with SAP SCM™, the two systems run in parallel for some time and the "hash totals" of both are compared. This comparison is used to fine-tune the new system, e.g., with respect to product classification, safety stocks, and days of demand coverage. Only when the SAP APO™ system provides at least the planning quality of previous solution is the cut-over to APO finalized.

5.4.4 SCM Based on SAP Systems

The main APO functionalities used by Hilti, DP and PP/DS plus deployment, are running on a daily basis. Because of the large data volume a powerful dedicated server is used. The current liveCache of the APO server amounts to 96 Gigabytes. Besides the enhanced functionality one reason for APO is to relieve the load on the transactional system. The SAP R/3™ system must provide online access almost around the clock, due to Hilti's global activities.

Stocks at the Hilti Centers are controlled via minimum and maximum inventory values. More than 100,000 stock keeping units (SKU) are defined at the plants and about 1,000,000 SKU at the marketing organizations. Hilti is able to retrieve information about all inventories at all SKU in real-time from a single system. Inventory management is highly automated: Stocks at the Hilti centers are refilled based on automatically generated orders to the warehouse. At the central warehouses purchase requisitions are generated automatically, and the material managers confirm these proposals in more than 90% of cases. This shows the high potential for automating routine decisions with sophisticated systems.

Initial approaches with CTM algorithms did not provide satisfactory results because CTM determines feasible solutions, for which all resources must be already available. However, in the manufacture of complex products with multi-stage

BOM it is typically not possible to wait until all parts are available, and production has to be started on the assumption that some parts that are still missing will nevertheless be provided in time.

Master data is transferred once a day via the standard integration functionality, the core interface (CIF). Transactional data is updated constantly in real-time between SAP R/3™ and SAP SCM™. The demand of the marketing organizations of each of the three main regions (Europe, Americas, Asia/Pacific) is planned individually.

The APO DP is based on an own BI, which is fed from the global BI system. The BI logic, which has influenced the design of APO DP, provides high flexibility in structuring the demand planning process. However, the necessity for detailed knowledge of both forecasting processes and BW technology makes implementation of the DP module anything but a trivial task. Hilti evaluated forecasting demand at the Hilti Centers level but dismissed this too-detailed approach because of poor forecast quality, unacceptable liveCache sizing, and long response times. Therefore, the demand forecasting units typically coincide with the warehouses where the data is more aggregated. The planners may adjust the forecast for such special events as promotions. Owing to the large number of products and storage points, applying advanced forecasting methods is not considered very important. The focus is on automating the process, because the large number of SKU cannot be handled manually by material managers.

The demand figures are transferred to the PP/DS system, which is used for generating purchase requisitions as well as planned orders (cf. Fig. 5.5). The planning horizon lies between 5 months and 18 months to allow adequate consideration of the lead times of materials and parts manufactured in house. The planning process is repeated each day from scratch. The data volume is huge – even just for the manufacturing plants, about 250,000 planned production and purchase requisitions result daily (performance is very important). PP/DS is used for infinite planning.

Only orders that will be executed in the next few days are considered for transfer into SAP R/3™ for execution. Because limited stock situations are not taken into account by PP/DS, it is necessary to apply the SNP deployment function. Quite surprisingly, executing deployment for the next few days takes longer than running the PP/DS application for scheduling of a far larger number of orders. Owing to its positive experiences with PP/DS, Hilti currently does not see any advantage to using SNP beyond the deployment functionality. PP/DS is also used by Hilti to support long-term decisions on capacity planning.

The Inventory Collaboration Hub (ICH) is not used by Hilti because ICH is more closely connected to R/3 and the existing link to APO is inadequate from the Hilti perspective. The purchase requisitions are not available in SAP R/3™ but only in APO, and therefore this information cannot be easily shared via the ICH. Hilti provides its main suppliers with information about orders, but also about expected demand for the planning horizon via Excel sheets. Much information provided by SAP Event Management (SAP EM™) is already available in the APO alert monitor, and thus far the company feels no need for additional information from SAP EM™.

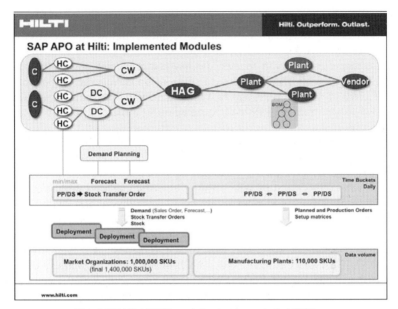

Fig. 5.5 SAP APO™ modules implemented at Hilti

5.4.5   Present Status and Future Developments

In spring 2008, Hilti was running SAP R/3™ 4.6c and SAP SCM™ 5.0 in its central IT department. All eight manufacturing plants and many marketing organizations were already using the central systems. About 250 users from 30 legal entities are authorized to access the SCM system; the number of concurrent SCM users may be as large as 150. The SCM roll-out to the remaining marketing organizations is expected to be finished by the end of 2008.

5.4.6   Interview

with Patrick Mayer, IT Teamlead for Supply Chain Planning, Hilti AG

CV

Patrick Mayer was born in Austria close to Lake Constance. After his engineering studies in electronics and telecommunications he also received his master's degree in Physics from the University of Innsbruck. Later a post-degree diploma in "International Management" at the University of Liechtenstein and UBC Vancouver (Canada) completed his studies. Starting in 1995, up to now Patrick has been working in different functions within the area of ERP and APS systems. Since joining Hilti in 1997 he has contributed to various international SCM-related projects within global IT. One of them was the well-known Hilti GPD/H2 initiative. Four years ago he assumed his current role as team and project leader for "Supply Chain Planning", focusing on SAP SCM™.

Gerhard Knolmayer: Which are the most business-relevant functionalities you can support by SAP SCM™ but not with SAP R/3™?

Patrick Mayer: The IT group simply could not support the global Supply Chain Management strategies of having a fully transparent stock, demand, and supply situation in one single system. It would not be possible to do a daily re-planning on the ERP for the whole Hilti supply chain, as the R/3 system is used inter-actively almost 24 h a day.

SCM demand planning is much more powerful, flexible, and customizable than any forecasting functionality in R/3. For instance, in DP it is easily possible to aggregate demand figures to any level, draw some conclusions, and read just immediately. PP/DS is able to plan across the network, leaving the single plant borders. This is crucial in our global environment. A traditional MRP system such as R/3 is not able to cover these requirements adequately.

Gerhard Knolmayer: What were the main reasons for replacing an earlier APS system by SAP APO™?

Patrick Mayer: There were many aspects: With the previous APS systems it is absolutely impossible to establish the expected close link to the SAP R/3™ system. Today's integration between SAP R/3™ and SAP SCM™ is very close and enables real-time synchronization for all transactional data.

The functional footprint of SAP SCM™ is quite big – that means we can address the various requirements in one system. Just think of the different expectations of a materials manager in one of the manufacturing facilities on the one hand and a market organization planner on the other.

Gerhard Knolmayer: What do you think are the critical success factors of implementing SAP SCM™ successfully?

Patrick Mayer: First of all there has to be a very good partnership between business and IT! Every company starting an SAP SCM™ implementation has to be very clear about the processes and functionalities that the system should cover. It is critical to understand which modules from this extensive APO suite have to be applied to get to the best possible solution. There are many modules covering the same or similar requirements, and at the beginning it is not very clear which way to go.

One key point is the correct sizing of the whole system (liveCache!). On the technology side it has to be considered that some know-how about liveCache is absolutely necessary. If DP is going to be used, the initial data design and some BI skills are crucial.

My recommendation would be to do a feasibility study with a reasonable amount of data to answer these questions. This will enable a faster and better realization afterwards.

5.5 Case Study Nestlé

5.5.1 Company Profile

Nestlé is the world's largest food and beverage company, founded and head-quartered in Vevey, Switzerland. It employs in excess of 250,000 people, has almost 500 factories situated in around 90 countries and its products are on sale everywhere. It has contracts with many co-manufacturers; some of whom work exclusively for Nestlé. In recent years, Nestlé has focused on becoming a nutrition, health and wellness company through its existing brands and new acquisitions.

The Nestlé brand portfolio covers practically all food and beverage categories: milk and dairy products, nutrition (infant, healthcare, performance and weight management), ice cream, breakfast cereals, coffee and beverages, culinary products (prepared dishes, cooking aids, sauces, etc.), chocolate, confectionery, pet care, and bottled water. Many brands have category leadership, both globally and in local markets. The best-known global brands include Nescafé, Nestea, Maggi, Buitoni, Purina, and, of course, Nestlé itself.

In 2007, Nestlé's revenues exceeded 107 billion Swiss Francs (about 100 billion USD at prevailing exchange rates).

Nestlé is run as a decentralized group of companies which foster and rely on organizational learning within the Group. Headquarters provides leadership, guidance, and consultancy to the regional and local organizations. Central competence units identify opportunities within Markets and Businesses and assist local management in improving business performance.

5.5.2 Supply Chain Management Organization

Almost 30,000 employees are involved in Nestlé's SCM, which is organized as follows:

- Corporate Operations – Supply Chain is one of the corporate functional units hosted at Headquarters and is responsible for
  - Global Supply Chain Strategy
  - Governance
  - Best Practices
  - Compliance
  - Market Assistance and Guidance, and
  - Reporting Standards and Guidelines.

- Zone Supply Chain is hosted by the regional line management organizations and is responsible for
  - Zone Strategy and Priorities
  - Regional Process Optimization
  - Regional Execution
  - Coordination, and
  - Reporting.
- Global Businesses Supply Chain is part of the Headquarters of the Nestlé businesses which are managed on a global basis (as opposed to those managed by the Zones) and is responsible for
  - Business Strategy and Priorities
  - Process Optimization
  - Coordination
  - Execution, and
  - Reporting.
- Region/Market Supply Chain is part of Nestlé operational companies and is responsible for
  - Region/Market Shared Services
  - Market/Country Priorities, and
  - Execution (Warehouses, Customer Service).

Nestlé uses a variety of Key Performance Indicators (KPIs), typically aligned with industry best practices. Global trends are reviewed monthly, and the Supply Chain Leadership team (including Corporate Supply Chain and the Supply Chain Heads of each Zone and Global Business) meets bi-monthly to coordinate activities cross-Zone and cross-Business. The diversity of SCM needs necessitates varied Supply Chain strategies, always driven by products, which are adapted to local tastes, market culture, and available infrastructure. For example, supply strategies utilize both dedicated local factories supplying a 400 km radius to global factories supplying multiple markets. The Supply Chain at Nestlé must manage all these complexities to define the most appropriate execution processes to meet customer and consumer requirements.

5.5.3   GLOBE – Global Business Excellence

GLOBE (Global Business Excellence) was launched in July 2000 and at the time was the most ambitious business process reengineering programme initiated by any multinational. Through the implementation of harmonized processes and best practices, common data standards and standardized information systems, the performance and operational efficiency of Nestlé businesses worldwide was targeted for improvement.

The rationale behind GLOBE was

- to leverage size as a strength in a rapidly changing environment,
- to unite and align on the inside so as to be more competitive on the outside, and
- to enable Nestlé to manage complexity with operational efficiency.

Positioned as a business project, GLOBE however has strong implications for information systems and information technology. Before GLOBE, hundreds of entities and factories were running systems and technology which evolved independently of one another. Determining relevant business data for accounting and decision making was a tedious process, as many systems were not interfaced. Thanks to GLOBE, Nestlé standardized both its business processes and the support of these processes. GLOBE consolidated numerous Data Centres into four, one for the Americas, one for Europe, one for Africa, Asia, and Oceania, and a fourth for global systems.

A central competence centre, the Business Technology Centre (BTC), was created to design and construct solutions on a global scale and to deliver them to three regional GLOBE Centres, responsible for implementing these solutions in the Markets and the Globally Managed Businesses (e.g., Nestlé Waters, Nestlé Nutrition).

By May 2008, GLOBE had been deployed in almost all of its operations around the world including factories, distribution centres, and sales offices. The opportunity for Nestlé now is to leverage GLOBE to deliver greater internal efficiencies.

SCM covers four key areas within GLOBE:

1. Procurement, from purchasing materials to paying suppliers. Emphasis is given to strategic procurement and a common procurement process for all goods and services across all companies has been established. Buying is organized globally or regionally, and locally only in the case of clear business benefits. Employees are empowered to perform operational level procurement, and improved information is available to support buyers in contract negotiations.
2. Customer Service consists of the order management process (including direct delivery) as well as processes related to claims, returns, and refusals.
3. Materials Handling, including physical logistics optimization, network optimization, transport cost minimization, and the provision of real-time data.
4. Demand and Supply Planning, covering strategic, tactical, and operational planning.

5.5.4 Technical Implications of GLOBE

Access to GLOBE systems is needed 24 h per day, 365 days per year, and with respect to planned and unplanned outages, a Service Level close to 100% needs to be and is provided. To balance functional requirements, technical constraints, and to manage risk, a split-system architecture was implemented for the GLOBE systems installed in all Data Centres. The information technology architecture of each Data Centre is very similar in order to facilitate change management, support, and disaster recovery planning.

Local, regional and in many cases global reporting is now available on a daily basis with Nestlé planning further enhancements and consolidation to make additional information available on a daily and/or real-time basis. Drill-down to detailed data of specific entities is also possible. More than 3 million sales orders are processed monthly through GLOBE and storage on the GLOBE systems has reached 2.65 Petabytes and grows daily. System storage management is a challenge with the deletion and archiving of data often being technically difficult due to complex interdependencies. In order to manage system storage, a Data Archiving Strategy has been established and is being implemented currently.

5.5.5   SCM and GLOBE Based on SAP Systems

Information system support of business processes is primarily standardized on SAP systems. Nestlé closely cooperates with SAP development on a number of projects, and there is a global contract between Nestlé and SAP for the worldwide use of SAP's software. At the end of 2007, Nestlé was running systems such as SAP ERP™ 4.7, SAP CRM™ 3.0, and also SAP SCM™ 4.1. Currently Nestlé are evaluating upgrades to SAP ECC™ 6.0, SAP CRM™ 7.0, and SAP SCM™ 6.0.

To meet Nestlé's requirements, some add-ons to the SAP systems have been developed in cooperation with SAP, and, where agreed, these add-ons will eventually be delivered as part of the standard SAP systems. Nestlé sometimes participates in the ramp-up phase of new releases to influence SAP's further development activities. Complementary systems are used to support areas where the present state of SAP systems does not provide required functionality. For example, Nestlé is using

- an eSupplyChain marketplace originally established by Nestlé, Danone, Henkel, and SAP, and later sold to Accenture,
- a Customer Payment Card from CyberSource,
- ARIBA for e-sourcing, and
- Software for co-managed inventory from INFLUE (downstream to customers).

GLOBE provides an extensive standardized set of systems and processes with accompanying tools, templates, and learning materials (referred to as the "GLOBE Solution"). The use of the GLOBE Solution can be adapted with respect to the environment in which individual operating companies/entities are acting (e.g., local legal and fiscal requirements). However, the majority of processes and procedures are standardized and enforced by headquarters; for instance, the demand planning process is driven through SAP APO™, although there are still some legacy tools in use in some Markets. Nestlé has also developed tools and additional algorithms which help the different entities in their use of SAP APO™. The associated upgrade risks from this bespoke development are understood and taken into consideration, with Nestlé trusting its strong information system/ technology know-how and experience to guide upgrade activities.

To improve data quality, a SAP Master Data Repository has been established. For example, a globally valid material number exists for each material. This results in better visibility and allows common purchasing activities and the negotiation of better purchasing conditions. The creation and validation of master data is time consuming and a constant challenge (cf. also Knolmayer and Röthlin 2006) but the resulting improvements has brought major benefits. One key lesson is that before harmonizing certain materials, e.g. raw materials, it is vitally important to harmonize material specifications.

Many products exist in several variants, taking local tastes into account. Master data, such as characteristics, classes, global charts of account, or cost elements held in the Master Data Repository, are sent to all regional systems. Data may be added in the different systems for local operations but the central master data must not be changed. This allows global and regional decision makers to obtain coherent, consolidated reporting.

5.5.6   Present Status (Mid 2008)

Nestlé is running SAP SCM™ 4.1 on two hardware systems in each Data Centre, with one system supporting Demand Planning (DP) and the other Supply Network Planning (SNP), Production Planning/Detailed Scheduling (PP/DS), and Global Available-to-Promise (ATP). The latter modules are interfaced via the Core Interface (CIF) with SAP R/3™ systems which also run on different hardware systems. All APO modules and the R/3 systems deliver data to the Business Warehouse (BW) to determine the KPIs at least on a weekly basis. DP is also receiving data from the BW.

Demand and Supply Planning is done on a planning horizon of 18 months, across which differing granularity is used. About 20% of the products are innovated or renovated across this horizon.

Demand Planning is used to determine the statistical baseline forecast. Promotional uplifts are interfaced from SAP CRM™ to provide a total demand forecast with promotions typically planned 6 months in advance. Repeat promotions can be statistically planned with sufficient accuracy. Problems result primarily from promotions which must be done within a very short time-frame.

SNP is used to support the distribution of final products. Thus far, the SNP heuristic has been used. The SNP Optimizer is considered an option that may become relevant in the future. For ease of use and to improve consistency, ABAP macros have been developed, e.g., for supporting Available-to-Deploy (ATD) decisions. Safety stocks are determined in SNP. In mid 2008, Nestlé was evaluating add-on tools to SNP to improve safety stock planning at both a strategic ("multi-echelon") and tactical ("site") level.

PP/DS is used in about 90% of factories, with different levels of sophistication for master production scheduling and detailed scheduling of final and semi-finished products. Raw and packaging materials are handled in SAP R/3™. Subcontracting is also controlled via PP/DS, but the functionality provided in APO for this process has proved somewhat crude to date.

ATP scope of check is generally applied to existing inventory and less so to planned receipts (i.e. production process orders). Problems are often solved through close collaboration with retailers. There is some reluctance to use the POS data of retailers because of the huge amount of data and the costs resulting from such agreements.

The stock transfer process within SAP systems is considered weak (e.g., within the Transport Load Builder, the algorithms are considered limited and it does not take account of vehicle utilization). As a result Nestlé custom add-ons were developed to meet business requirements.

Nestlé customized the CIF to adapt it to its split-system architecture with CIF appends and user exits extensively used. Minor changes to the CIF can have dramatic effects on the performance of the systems. The volumes passing through the CIF at Nestlé are unprecedented and any CIF backlogs can rapidly cause execution problems, with much emphasis placed on ensuring problem free operation of the CIF. Many reports had to be adapted to take into account the split-system architecture.

5.5.7   Future Developments

Some of the other major areas of development being undertaken are

- TRM (Task & Resource Management): The ability to optimize movements inside a warehouse based on the task priority and location of a warehouse operator.
- Dock & Yard Management: The control of both the Warehouse dock doors and the yard to enable planning and control over these functions and integration with both receipts and dispatches.
- Management Dashboards: Tactical reporting/alerting by proactive monitoring of processes, including Procurement, Factory and Warehouse Operations, and Order Management.
- Collaboration (Vendor and Customer): Various initiatives to link Trading Partners via Web Portal functionality. These include Co-Manufacturing/ Subcontracting operations, B2B portal, Forecast Collaboration, and Document Exchange (e.g., with respect to specifications).

5.5.8   Interview

with Tony Borg, Vice President Corporate Operations, Supply Chain, Nestlé SA

CV

Tony Borg is an Australian of Maltese origin. He completed most of his studies in Europe before emigrating to Australia in 1973 both to complete his studies and to pursue a career in Finance. He joined Nestlé in Finance and over subsequent years was exposed to all facets of the business, working at different sites including factories. When Supply Chain started taking on a new meaning in the mid 1980s, Tony assumed various responsibilities in Supply Chain ending up as Head of Supply Chain Development and Logistics for the Region based in Australia. During his latter years Tony was seconded to L'Oréal Australia for 3 years as Head of Supply Chain and ISIT (Information Systems and Information Technology). In 2003 he was transferred to the Group's headquarters in Vevey, Switzerland, to join the GLOBE project, heading Business Excellence for Supply Chain. After 3 years in GLOBE he assumed his current role as Head of Supply Chain for the entire Nestlé Group.

Gerhard Knolmayer: Nestlé is a highly diversified company with a broad spectrum of products and acting in very different markets. What are the main advantages in standardizing processes and IT support in such a diverse environment?

Tony Borg: The Company has been around for more than 140 years, since well before modern technology capabilities. The markets therefore went through a natural technology evolution and exploited technology to enable ongoing competitiveness. It stood to reason therefore that very few markets had a common infrastructure and common ways of working. Moreover, the data standards, to a large extent, evolved within each market and sometimes even within individual sites; as more global visibility was requested, consolidation of information was very difficult and extremely laborious at corporate level. The intention of GLOBE was to bring "order" into our company and enable effective management of increasing complexity while leveraging scale. The program was founded on three objectives:

- Common processes through defined Best Practices,
- Common data standards, and
- Common IT platform.

The GLOBE Solution has now provided us with a base to share ways of working more easily and thus leverage learnings from one part of the world to the other. The standardization of the data makes internal consolidation of information easier. It also facilitates external data exchange with our trading partners. Lastly the IT infrastructure facilitates the coordination of the large volume of data flows between the different systems and provides us with better redundancy of hardware.

Gerhard Knolmayer: At some phases of the GLOBE project the local management showed reluctance to this standardization approach. How is GLOBE seen today in Nestlé and what have been the biggest benefits of GLOBE?

Tony Borg: The GLOBE vision was initiated at Group level through our then CEO Mr. Peter Brabeck-Letmathe. It represented a sound long-term vision and strong conviction to conduct such a colossal change when the Company was not in a crisis. It was natural for some apprehension to creep in bringing the old adage "why fix something that is not broken?" Management in the Markets however recognized that with our increasing size and complexity GLOBE was a must. Initial resistance was overcome through implementation successes and ongoing communication. The Company culture is such that once the goal is clear we actually rally and run faster and we were able to implement in 5 years what a lot of other Companies failed to achieve.

Now that we have pretty much the total Group on GLOBE, the Markets and indeed the whole Business is seeing the opportunities through procurement for instance. Another clear benefit is that we no longer implement a tool or best practice on a Market-by-Market basis. Once we implement a tool or new functionality it becomes available to the whole Group not just to Markets big enough to afford it.

Gerhard Knolmayer: How much is the Corporate Operations unit using APO directly?

Tony Borg: APO is our key planning tool. The sophistication of usage varies between Markets, and frankly we continue learning about how best to optimize our planning through this tool. We are working with our GLOBE Centers to improve the skills of our people to be able to better exploit the functionality available and hence improve our overall planning across the longer term horizon.

Chapter 6

Supply Chain Management in Midsize Companies, Based on the New Hosted Solution SAP Business ByDesign™

6.1   Cross Application Characteristics of SCM

Business ByDesign is a comprehensive software solution for midsize companies that covers all essential fields of enterprise applications. Besides Supply Chain Management (SCM) Business ByDesign contains the following applications:

- Executive Management Support
- Financial Management
- Customer Relationship Management
- Supplier Relationship Management
- Human Resources Management
- Project Management
- Compliance Management.

Also in regard to functional depth diverse types of functionality are consistently represented in one application: Business ByDesign comprises functionality which is usually spread to specialized application types such as ERP Systems, Advanced Planning Systems (APS), or Business Intelligence (BI).

As shown in Fig. 6.1, SAP Business ByDesign™ is an application arranged according to the concept of Service-oriented Architecture. A distinction is made in all applications between the User Interface Level and the Application Level lying beneath (Application Platform) which manages the tasks of modeling, data handling, and business process implementation. This architecture improves the application's standardization on each of these levels. Hence, the system facilitates the daily work of typical midsize company users who are rather all-rounders than specialized in single functions.

The implementation's starting points are the Business Objects of the Application Platform. A Business Object defines, for example, a master data object, an order or a plan regarding its structure as well as the functions executable within the structure.

In a functional view the Business Objects within the Application Platform can be grouped into Process Components and these in turn into applications. In a process-related view processes can be seen as messages or interactions between

Business Objects across process components and applications. These processes are linked to Business Scenarios such as "Sell from stock" or "Third-party direct ship." Such business scenarios mark complete and consistently configured process tracks along several applications.

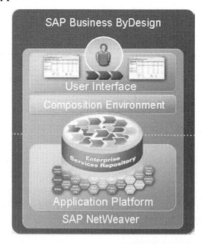

Fig. 6.1 Business ByDesign as a platform-based solution

The term Business ByDesign points out that all Business Objects, the whole functionality, and all business scenarios combined out of it are "model-based." In other words, it is about standardized software which is ready for use within a short period of time with few steps of configuration and parameterization. Because of the "model-based approach" conceivable options mark a closed solution space. Thus, the configuration of the system can be deduced by means of a set of rules from answers to a few initial questions (Adaptation Catalogue). Therefore, detailed implementation concepts, extensive customizing, and substantial on-site tests can be avoided. Instead the configuration is done in two steps, which are called Business Configuration and Fine Tuning.

Besides fast system implementation SAP Business ByDesign™ qualifies as mid-size company-friendly system because it can be operated as a hosted solution. The expenses connected with provision of hardware, software maintenance, and oper-ations are reduced.

Another useful part of the Business ByDesign solution for the midsize company user is the so-called Business Task Management. Thereby it is possible to generate rule- or exception-based task propositions along the Business Scenarios. It is not only a matter of warnings that enter the task list of the affected user, but also of guiding the user to appropriate subsequent activities, such as an approval.

The decoupling of software layers allows improving the user interface across applications in regard to user-friendliness. Overall concepts of the Business By Design user interface are also of benefit to the application SCM. These concepts include "Work Centers" (Menu Structure), "Role Based Access Management" (Authorization), and "Workplace Embedded Knowledge Transfer" (Learning Environment).

6.2   Components of SCM in SAP Business ByDesign™

Fig. 6.2 shows the components within the application SCM without references to adjoining applications, such as Customer Relationship Management.

The process of planning begins in the component Demand Planning which serves to forecast demand. Statistical forecast data generated by the system can be adapted manually before being released to Supply Chain Planning and Control. Thereby, supply can be planned well in advance of actual sales orders.

As the first step in Supply Chain Planning and Control (SCPC) forecast is matched with actual demand in order to determine total demand. Finding the source of supply for external demand and the availability check also take place in SCPC. The main step of SCPC is to generate production or purchasing proposals according to total demand. Corresponding requisitions are released afterwards to Manufacturing Execution and/or Supplier Relationship Management.

The detailed planning and control level of SCM is constituted by Warehousing and Manufacturing Execution. While Manufacturing Execution represents the company's production system, Warehousing Execution handles the enterprise logistics. Warehousing and Manufacturing Execution is characterized by a scalable representation of resources and a consistent technical integration with machinery control systems, mobile devices plus tracking and identification systems of the real world.

Fig. 6.2 Components of the application SCM in SAP Business ByDesign™

6.3   SCM Principles in SAP Business ByDesign™

Besides the elementary design principles across all applications we have to refer here in particular to the principles within SCM. These are:

- Demand Driven Planning and Execution
- Decoupling of Planning and Execution
- Interlocking of Warehousing and Manufacturing Execution
- Supporting Execution Level Users.

6.3.1   Demand Driven Planning and Execution

The fundamental idea of SCM for SAP Business ByDesign™ is that the competitive advantage of a midsize company lies in its flexibility in reacting on changed customer needs in qualitative, quantitative, and structural respects.

The use of software should not pose additional challenges for a midsize company if demand rises or falls unexpectedly or an elementary shift of the product mixes – required by the market – takes place. On the contrary, the software has to be a pragmatic support under such circumstances.

The business scenario "Make to Specification" is an example how SCM in SAP Business ByDesign™ addresses this kind of requirements. In this scenario the individual order can be represented by using standardized products. A manufacturer who produces windows with customer-specific measurements out of standardized dimensions can thus plan order-neutral and likewise administer individual orders. The user can decide whether to assign recurrent product characteristics (e.g., color) to the product. If more appropriate the user also can handle specifications informally in the form of text attached to the product used in the order (e.g., order-specific measurements of the product).

One more risk of system-immanent rigidity lies in the use of decision models in production planning. To choose appropriate decision models and to have to parameterize these each time accurately means a hurdle to midsize enterprises. For this reason SAP Business ByDesign™ avoids sophisticatedly parameterized optimizers as used in APS systems. The planning and controlling system is rather a "what-if" system which shows the consequences of changed planning situations through realistic capacitive modeling, alerts management, and integrated reporting.

Demand driven planning and execution relies on the user's ability to judge and react. Therefore, the containment of variance in cycle time is particularly important. SCM in SAP Business ByDesign™ accounts for the reduction and stability of cycle times in many ways, e.g., taking account of capacities in planning, adaptive inventory planning, anticipatory capacity planning, and reporting of cycle times. The execution level features of close interlocking of warehousing and manufacturing, unified task management, and real world integration – explained in more detail below – are also closely related to the reduction of cycle times.

6.3.2 Decoupling of Planning and Execution

As indicated in Section 6.2, the dividing line between production planning and execution in SAP Business ByDesign™ is drawn such that the detailed production planning is allocated together with production control to Manufacturing Execution. Thus, a better interlocking of detailed planning with the real situation in the internal logistics is achieved. The rough-cut planning realized in Supply Chain Planning and Control helps on the other hand to get a realistic picture for the overall co-ordination of production.

Rough-cut planning primarily has the function of smoothing the level of intensity in Manufacturing and Warehousing Execution by getting adaptations under way in time. This affects mainly the adaptation of inventory postponement policies for components and the determination of capacity adaptations.

The features of demand and supply planning expressed in the term "Advanced Planning" cover the infinite planning of supply combined with the display of capacity utilization, leveling of capacity, multi-level pegging between supply and demand, the generation of alerts, etc.

6.3.3 Interlocking of Warehousing and Manufacturing Execution

Demand driven supply planning also requires a continuous flow of material and information across the supply chain between supplier, internal supply chain, and customer at the execution level. As a consequence the areas of Manufacturing Execution and Warehousing Execution are interlocked in Business ByDesign.

As Fig. 6.3 shows, SAP Business ByDesign™ accounts for the frequent transfers in the flow of orders between Manufacturing Execution and Warehousing Execution. The Business Objects of the execution level are modeled according to the sequence of activities and transfers between quality control, production, and warehousing. The processes of inbound and outbound logistics are modeled symmetrically as far as possible. The tasks of quality control are designed consistently, be they located in Warehousing Execution or in Manufacturing Execution.

Fig. 6.3 Key areas of warehousing and manufacturing execution in SAP Business ByDesign™

6.3.4   *Supporting Execution Level Users*

In most of today's enterprise software implementations there is a line of division between core ERP systems and execution level legacy systems in Warehousing and Manufacturing. As opposed to this practice Warehousing and Manufacturing Execution is an essential part of SCM in SAP Business ByDesign™. Therefore this component is designed to become the preferred workplace of the blue collar worker, where he can process his work items or tasks, such as pick, put away, goods receipt, or rework. The feature of "Logistics Task Management" creates links between tasks, the assigned employees and the enterprise resources. It is designed as a unified point of entry to execute diverse types of tasks. Equipment used in the real world is integrated into Logistics Task Management; for example, tasks can be selected and executed remotely from the menu of a mobile device. Warehousing and Manufacturing Execution is designed such that it can easily take account of interfaces to emerging technologies as well. For instance, integration of Radio Frequency Identification (RFID) could be considered as a feature that might be available in future releases.

6.4   Summary and Prospects

The application SCM in SAP Business ByDesign™ is aligned to the needs of midsize companies in both a cross-application view and an SCM-specific view. As SAP Business ByDesign™ was launched relatively recently, the number of business scenarios available will grow. Companies which may apply business scenarios that are already available obtain an affordable product that can easily be adopted. The product emphasizes the principles of a decentralized, individualized planning and execution, in order to facilitate adaptation to changes in midsize companies' markets. The roadmap for diversification of the catalogue of available business scenarios will predominantly consider customer feedback from the midsize company sector.

Abbreviations

3PL	Third-Party Logistics Provider
4PL	Fourth-Party Logistics Provider
ABAP	Advanced Business Application Programming
AII	Auto-ID Infrastructure
ALE	Application Link Enabling
APO	Advanced Planning and Optimization
APQC	American Productivity & Quality Center
APS	Advanced Planning and Scheduling; Advanced Planning System
ARIS	Architecture of Information Systems
ASN	Advanced Shipping Notifications
ASP	Application Service Providing
ATD	Available-to-Deploy
ATP	Available-to-Promise
B2B	Business-to-Business
BI	Business Intelligence
BOD	Bill of Distribution
BOM	Bill of Materials
BOP	Backorder Processing
BW	Business Information Warehouse
CAD	Computer Aided Design
CBF	Characteristic-Based Forecasting
CCOR	Customer Chain Operations Reference
CEO	Chief Executive Officer
CIF	Core Interface
CMDS	Collaborative Management of Delivery Schedules
COL	Supply Chain Collaboration
CPFR	Collaborative Planning, Forecasting, and Replenishment
CRM	Customer Relationship Management
CSU	Complex Shipping Unit
CTM	Capable-to-Match
CTP	Capable-to-Promise
CUA	Central User Administration
CV	Curriculum Vitae
CW	Central Warehouses
DC	Distribution Center
DCOR	Design Chain Operations Reference
DEPL	Deployment and Inventory Balancing

DIF	Demand-Influencing Factors
DIMP	Discrete Industries and Mill Products
DP	Demand Planning
DRP	Distribution Requirements Planning
DS	Detailed Scheduling
DVD	Digital Versatile Disc
EBPP	Electronic Bill Presentment and Payment
ECM	Engineering Change Management
EDI	Electronic Data Interchange
EM	Event Management
EOQ	Economic Order Quantity
EP	Enterprise Portal
EPC	Electronic Product Code
ERP	Enterprise Resource Planning
et al.	et alii (and others)
EWM	Extended Warehouse Management
F&R	Forecasting & Replenishment
FCS	Forecasting
GATP	Global Available-to-Promise
GLOBE	Global Business Excellence
GMP	Good Manufacturing Practices
GPD	Globalization of Processes and Data
GTS	Global Trade System
GUI	Graphical User Interface
HAG	Hilti AG
HC	Hilti Centers
HR	Human Resources
HU	Handling Unit
ICH	Inventory Collaboration Hub
ID	Identification
IDN	Inbound Delivery Notification
IDOC	Intermediate Document
INVP	Inventory Planning
IS	Information Systems
ISIT	Information Systems and Information Technology
IT	Information Technology
JIT	Just-in-Time
KPI	Key Performance Indicators
LES	Logistics Execution System
LIS	Logistics Information System
MbE	Management by Exception

MISL	Multi-Item Single Delivery Location
MLR	Multiple Linear Regression
MM	Material Management
MMP	Model-Mix Planning
MO	Market Organizations
MRO	Maintenance, Repair, and Overhaul
MRP II	Manufacturing Resource Planning
MSP	Maintenance and Service Planning
n.a.	not available
ODL	Order Due List
ODM	Order Data Management
OEM	Original Equipment Manufacturer
OER	Object Event Repository
OLTP	Online Transaction Processing
PDA	Personal Digital Assistant
PDM	Product Data Management
PDS	Production Data Structure
PFS	Process Flow Scheduler
PI/XI	Process Integration/Exchange Infrastructure
PLM	Product Lifecycle Management
PML	Physical Markup Language
POD	Pending Obsolescence Date
POS	Point of Sales
PP	Production Planning
PP/DS	Production Planning and Detailed Scheduling
PPM	Production Process Model
PP-PI	Product Planning - Process Industries
qRFC	Queued Remote Function Calls
R&D	Research & Development
RFC	Remote Function Calls
RFID	Radio Frequency Identification
ROCE	Return on Capital Employed
ROI	Return on Investment
RR	Responsive Replenishment
RTI	Returnable Transport Items
SC	Supply Chain
SCC	Supply Chain Council
SCEM	Supply Chain Event Management
SCM	Supply Chain Management
SCOR	Supply Chain Operations Reference
SCPC	Supply Chain Planning and Control

SD	Sales and Distribution
SED	Stock Exhaustion Date
SEM	Strategic Enterprise Management
SKU	Stock Keeping Unit
SME	Small and Medium Enterprises
SMI	Supplier-Managed Inventory
SMS	Short Message Service
SNC	Supply Network Collaboration
SNP	Supply Network Planning
SOB	Surplus and Obsolescence Planning
SPE	Service Parts Execution
SPM	Service Parts Management
SPP	Service Parts Planning
SPPD	Successor Product Planning Date
SPRD	Successor Product Receipt Date
SRM	Supplier Relationship Management
SUP	Supplier Collaboration; Supersession
TLB	Transport Load Builder; Transport Load Building
TP/VS	Transportation Planning and Vehicle Scheduling
TRM	Task & Resource Management
TSDM	Time Series Data Management
TSL	Target Service Levels
TSP	Transportation Service Providers
TV	Television
USD	US Dollar; $
VICS	Voluntary Interindustry Commerce Solutions
VLCO	Virtual Location for Consolidated Ordering
VMI	Vendor Managed Inventory
w/o	without
WIP	Work in Progress
XBRL	Extensible Business Reporting Language
XI	Exchange Infrastructure
XML	Extensible Markup Language

References

Ackermann, I. (2003) Using the Balanced Scorecard for Supply Chain Management – Prerequisites, Integration Issues, and Performance Measures. In: Seuring, S., Müller, M., Goldbach, M., Schneidewind, U. (Eds.): Strategy and Organization in Supply Chains. Heidelberg/New York: Physica, 289-304.

Berning, G., Brandenburg, M., Gürsoy, K., Mehta, V., Tölle, F.-J. (2002) An integrated system solution for supply chain optimization in the chemical process industry. OR Spectrum 24/4, 371-401.

Bolstorff, P. (2006) Balancing Your Value Chain Metrics. http://www.scelimited.com/sitebuildercontent/sitebuilderfiles/balancingyourvaluechainmetrics.pdf

Boyson, S., Harrington, L.H., Corsi, T.M. (2003) In Real Time – Managing the New Supply Chain. Westport/London: Praeger.

Brauchle, T. (2006) Supplier Collaboration to Reduce Stock Levels. Presentation at SAPPHIRE'06, Paris.

Bretzke, W.-R. (2006) SCM: Sieben Thesen zur zukünftigen Entwicklung logistischer Netzwerke. Supply Chain Management 6/3, 7-15.

Butner, K. (2007) Blueprint for supply chain visibility. IBM Global Business Services, G510-6644-00, Somers, 1-2. http://www-935.ibm.com/services/us/gbs/bus/pdf/g510-6644-00-scblueprint.pdf

Camerinelli, E., Cantu´, A. (2006) Measuring the Value of the Supply Chain: A Framework. Supply Chain Practice 8/2, 40-59.

Chatfield, D.C., Kim, J.G., Harrison, T.P., Hayya, J.C. (2004) The bullwhip effect - impact of stochastic lead time, information quality, and information sharing: A simulation study. Production and Operations Management 13/4, 340-353.

Chatterjee, D. (2001) Capturing Value Through SCOR Metrics. Presentation at SAP Asia 2001. http://www.supply-chain.org/site/files/Chatterjee_SAP_SCWSEA02.zip

Chen, Z.-L., Vairaktarakis, G. (2005) Integrated Scheduling of Production and Distribution Operations. Management Science 51/4, 614-628.

de Souza, R., Khong, H.P. (1999) Supply chain models in hard disk manufacturing. IEEE Transactions on Magnetics 35/2, 950-955.

Dickersbach, J.T. (2005) Characteristic Based Planning with mySAP SCM^{TM}. Berlin/Heidelberg: Springer.

Dickersbach, J.T. (2006) Supply Chain Management with APO. Structures, Modelling Approaches and Implementation of mySAP SCM 4.1. 2nd ed., Berlin/Heidelberg: Springer.

Dickersbach, J.T. (2007) Service Parts Planning with mySAP SCM^{TM}. Processes, Structures, and Functions. Berlin/Heidelberg: Springer.

Dießner, P., Rosemann, M. (2007) Supply Chain Event Management: Managing Risk by Creating Visibility. In: Ijioui, R., Emmerich, H., Ceyp, M. (Eds.): Strategies and Tactics in Supply Chain Event Management. Berlin/Heidelberg: Springer, 83-98.

Disney, S. M., Towill, D. R. (2003a): The Effect of Vendor Managed Inventory (VMI) Dynamics on the Bullwhip Effect in Supply Chains. International Journal of Production Economics 85/2, 199-215.

Disney, S. M., Towill, D. R. (2003b): Vendor-Managed Inventory and Bullwhip Reduction in a Two-Level Supply Chain. International Journal of Operations & Production Management 23/6, 625-651.

Dudek, G. (2004) Collaborative Planning in Supply Chains. Berlin et al.: Springer.

Dudek, G., Stadtler, H. (2005) Negotiation-based collaborative planning between supply chains partners. European Journal of Operational Research 163/3, 668-687.

Eßig, M. (2006) Supply Chain Management: Idealtypisches Paradigma oder realistisches Konzept? Supply Chain Management 6/11, 55-56.

Fleischmann, B., Meyr, H. (2003) Planning Hierarchy, Modeling and Advanced Planning Systems. In: de Kok, A.G., Graves, S.C. (Eds.): Supply Chain Management: Design, Coordination and Operation. Amsterdam et al.: Elsevier, 457-523.

Fleischmann, B., Meyr, H., Wagner, M. (2005) Advanced Planning. In: Stadtler, H., Kilger, C. (Eds.): Supply Chain Management and Advanced Planning: Concepts, Models, Software and Case Studies. 3rd ed., Berlin et al.: Springer, 81-106.

Forrester, J.W. (1961) Industrial Dynamics – a major breakthrough for decision makers. Harvard Business Review 36/4, 37-46.

Gassmann, M. (2001) Incorporating SCOR Metrics into SCM Software Products. Presentation at SCC World, Berlin. http://www.supply-chain.org/site/files/Gassmann_SAP_SCWE01.zip

Gould, L. (2005) What You Need To Know About Supply Chain Management. Automotive Design and Production, No. 2. http://www.autofieldguide.com/articles/020511.html

Hausman, W.H. (2004) Supply Chain Performance Metrics. In: Harrison, T.P., Lee, H.L., Neale, J.J. (Eds.): The Practice of Supply Chain Management: Where Theory and Application Converge. New York: Springer, 61-73.

Hendricks, K.B., Singhal, V.R. (2003) The effect of supply chain glitches on shareholder wealth. Journal of Operations Management 21/5, 501-522.

Hendricks, K.B., Singhal, V.R., Stratman, J.K. (2007) The impact of enterprise systems on corporate performance: A study of ERP, SCM, and CRM system implementations. Journal of Operations Management 25/1, 65-82.

Hieber, R. (2002) Supply Chain Management: A Collaborative Performance Management Approach. Zürich/Singen: vdf Hochschulverlag.

Huan, S.H., Sheoran, S.K., Wang, G. (2004) A review and analysis of supply chain operations reference (SCOR) model. Supply Chain Management 9/1, 23-29.

IDS Scheer (2005) Business Process Excellence for Supply Chain Optimization, White Paper, Saarbrücken. http://wp.bitpipe.com/resource/org_1111097862_533/SCM_APO_edp.pdf?site_cd=fbs

IDS Scheer (2007) ARIS EasySCOR. http://www2.ids-scheer.com/us/products/aris-easyscor.htm

Intentia (2001) Continuous Replenishment Program & Vendor Managed Inventory. http://www.vendormanagedinventory.com/crp.pdf

Jia, H.Z., Fuh, J.Y.H., Nee, A.Y.C., Zhang, Y.F. (2002) Web-based Multi-functional Scheduling System for a Distributed Manufacturing Environment. Concurrent Engineering 10/1, 27-39.

Kallrath, J., Maindl, T.I. (2006) Real Optimization with SAP APO. Berlin et al.: Springer.

Kaplan, R.S., Norton, D.P. (1996) The Balanced Scorecard. Boston: Harvard Business School Press.

Kleijnen, J.P.C., Smits, M.T. (2003) Performance metrics in supply chain management. Journal of the Operational Research Society 54/5, 1-8.

Knolmayer, G.F., Dedopoulos, I. (2006) Neugestaltung des Lademittel–Managements der Migros auf Basis von ASP. Volume 13, Case Studies of the Akademische Partnerschaft ECR Deutschland, Cologne. http://www.ecracademics.de/neugestaltung_des_lademittel-managements_der.php

Knolmayer, G.F., Röthlin, M. (2006) Quality of Material Master Data and Its Effect on the Usefulness of Distributed ERP Systems. In: Roddick, J.F. et al. (Eds.): Advances in Conceptual Modeling – Theory and Practice. Berlin: Springer, 362-371.

Lee, H.L., Amaral, J. (2002) Continuous and Sustainable Improvement Through Supply Chain Performance Management. http://jobfunctions.bnet.com/whitepaper.aspx?docid=72917

Lee, H.L., Padmanabhan, V., Whang, S. (1997a) Information distortion in a supply chain: The bullwhip effect. Management Science 43/4, 546-558.

Lee, H.L., Padmanabhan, V., Whang, S. (1997b) The bullwhip effect in supply chains. Sloan Management Review 38/3, 93-102.

Leitz, A. (2005) Supplier Collaboration mit dem mySAP SCM Inventory Collaboration Hub. Bonn: Galileo Press.

Mertens, P. (2007) Integrierte Informationsverarbeitung 1, Operative Systeme in der Industrie. 16th ed., Wiesbaden: Gabler.

Meyr, H., Wagner, M., Rohde, J. (2005) Structure of Advanced Planning Systems. In: Stadtler, H., Kilger, C. (Eds.): Supply Chain Management and Advanced Planning: Concepts, Models, Software and Case Studies. 3rd ed., Berlin/Heidelberg: Springer, 109-115.

Nestlé (2006) The world of Nestlé. http://www.nestle.com/Resource.axd?Id= 602C42FE-04D6-4669-BEE1-1027492FE5E8

Neumann, K., Schwindt, C., Trautmann, N. (2002) Advanced production scheduling for batch plants in process industries. OR Spectrum 24/3, 251-279.

Otto, A., Kotzab, H. (2003) Does supply chain management really pay? Six perspectives to measure the performance of managing a supply chain. European Journal of Operational Research 144/2, 306-320.

Plenert, G. (2007) Reinventing Lean – Introducing Lean Management into the Supply Chain. Amsterdam et al.: Elsevier.

Pohlen, T.L., Goldsby, T.J. (2003) VMI and SMI programs. How economic value added can help sell the change. International Journal of Physical Distribution & Logistics Management 33/7, 565-581.

Poluha, R.G. (2007) Application of the SCOR Model in Supply Chain Management. Youngstown: Cambria Press.

Roussel, J., Skov, D. (2007) European Supply Chain Trends 2006: Using the Supply Chain to Drive Operational Innovation. http://www.supply-chain.org/galleries/taxonomy/ PRTM%20Supply%20Chain%20Trends%20Report%202006.pdf

SAP (2000) mySAP Supply Chain Management at Cerveceria Nacional. http://www.sap.com/ industries/consumer/pdf/CSS_Cerveceria_National.pdf

SAP (2001) mySAP Financials Financial Operations: Electronic Bill Presentment and Payment (EBPP). http://www.sap.com/industries/publicsector/pdf/50046563.pdf

SAP (2007) Vendor Managed Inventory (VMI). http://www50.sap.com/businessmaps/ 459B5C21A96411D3870B0000E820132C.htm

Schmidt, R.S. (2007) Impact of Information Sharing and Order Aggregation Strategies on Supply Chain Performance. In: Proceedings of the 5th International Conference on Supply Chain Management and Information Systems, Melbourne.

Schmidt, R., Knolmayer, G.F. (2006) Simulationsstudien zu Information Sharing und Vendor Managed Inventory: Ein Vergleich. In: Wenzel, S. (Ed.): Simulation in Produktion und Logistik 2006. San Diego/Erlangen: SCS Publishing House, 93-102.

Schnetzler, M.J., Schönsleben, P. (2007) The contribution and role of information management in supply chains: a decomposition-based approach. Production Planning & Control 18/6, 497-513.

Simchi-Levi, D., Kaminsky, P., Simchi-Levi, E. (2008) Designing and Managing the Supply Chain. 3rd ed., Boston et al.: McGraw-Hill/Irwin.

Singh, M.P. (1999) The end of the supply chain? IEEE Internet Computing 3/6, 4-5.

Singhal, V.R. (2003) Quantifying the Impact of Supply Chain Glitches on Shareholder Value. SAP White Paper, Walldorf. http://www.sap.com/solutions/business-suite/scm/pdf/BWP_Quantify.pdf

Småros, J., Lehtonen, J.-M., Appelqvist, P., Holmström, J. (2003) The Impact of Increasing Demand Visibility on Production and Inventory Control Efficiency. International Journal of Physical Distribution & Logistics Management 33/4, 336-354.

Steinke, S. (2006) A380 cable problems threaten Airbus. Flugrevue No. 12, 22.

Sterman, J.D. (1989) Modeling management behavior: Misperceptions of feed back in a dynamic decision making experiment. Management Science 35/3, 321-339.

Supply-Chain Council (2007) SCORmark Benchmarking Portal. http://www.supply-chain.org/cs/benchmarking.

Supply-Chain Council (2008) Supply-Chain Operations Reference-model. SCOR Overview, Version 9.0. http://www.supply-chain.org/galleries/public-gallery/SCOR%209.0%20Overview%20Booklet.pdf

Supply Chain Process Improvement (2007) SCOR Metrics. http://www.scpiteam.com/SCOR%20Metrics.htm

The Performance Measurement Group (2006) The High Performance Supply Chain Report. 2006 Annual Report 1/1, Waltham. http://www.pmgbenchmarking.com/public/news/SCEP%20Exec%20Sum%20Annual%20Report%202006.pdf

The Performance Measurement Group (2007) Supply Chain Performance Indicator. http://www.pmgbenchmarking.com/public/product/scorecard/supply_chain/scorecard.asp

VICS (2004) CPFR: An Overview. http://www.vics.org/committees/cpfr/CPFR_Overview_US-A4.pdf

Windischer, A., Grote, G. (2003) Success Factors for Collaborative Planning. Seuring, S., Müller, M., Goldbach, M., Schneidewind, U. (Eds.): Strategy and Organization in Supply Chains. Heidelberg/New York: Physica, 131-146.

Wood, D.C. (2007) SAP SCM: Applications and Modeling for Supply Chain Management (with BW Primer). Hoboken: Wiley.

Index

Printing: Krips bv, Meppel, The Netherlands
Binding: Stürtz, Würzburg, Germany